Ubiquitous and Transparent Security

In an increasingly interconnected digital realm, *Ubiquitous and Transparent Security: Challenges and Applications* emerges as a guiding beacon through the intricate web of modern cybersecurity. This comprehensive tome meticulously dissects the multifaceted challenges faced in safeguarding our digital infrastructure. From the omnipresence of threats to the evolving landscape of vulnerabilities, this book navigates the complexities with a keen eye, offering a panoramic view of the security terrain.

Drawing on a rich tapestry of insights from leading experts, this book transcends the traditional boundaries of security discourse. It unveils innovative strategies and technologies, illuminating the path toward a future where security seamlessly integrates with the fabric of our digital existence. With a keen focus on transparency, it delves deep into the mechanisms that enable a clear, holistic view of security, empowering stakeholders to navigate this dynamic landscape with confidence.

From cutting-edge applications to the ethical considerations of ubiquitous security, each chapter acts as a guiding compass, providing actionable insights and fostering a deeper understanding of the intricate balance between accessibility and protection. *Ubiquitous and Transparent Security* is not merely a book; it's a roadmap for practitioners, policymakers, and enthusiasts alike, navigating the ever-evolving world of cybersecurity.

Each chapter within this compendium illuminates the diverse challenges that confront security practitioners, policymakers, and technologists today. It goes beyond the conventional paradigms, exploring the nuanced intersections between accessibility, transparency, and robust protection. Through a rich amalgamation of research-backed insights, real-world case studies, and visionary forecasts, this book offers a holistic understanding of the evolving threat landscape, empowering stakeholders to fortify their defences proactively.

Ubiquitous and Transparent Security

Challenges and Applications

Edited by A. Suresh Kumar,
Rajesh Kumar Dhanaraj, Yassine Maleh,
and Daniel Arockiam

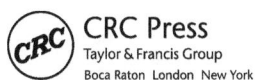

CRC Press
Taylor & Francis Group
Boca Raton London New York

CRC Press is an imprint of the
Taylor & Francis Group, an **informa** business

Designed cover image: © Shutterstock

First edition published 2024
by CRC Press
2385 NW Executive Center Drive, Suite 320, Boca Raton FL 33431

and by CRC Press
4 Park Square, Milton Park, Abingdon, Oxon, OX14 4RN

CRC Press is an imprint of Taylor & Francis Group, LLC

ISBN: 9781032424071 (hbk)
ISBN: 9781032424118 (pbk)
ISBN: 9781003362685 (ebk)

DOI: 10.1201/ 9781003362685

Typeset in Sabon
by Newgen Publishing UK

Contents

About the editors

A. Suresh Kumar graduated in Computer Science and Engineering (B.E) from Periyar University, Salem in 2003. He completed his Master's (M.E) in Computer Science and Engineering from Anna University, Trichy in 2010. He was awarded a Ph.D. in Computer Science and Engineering from Anna University, Chennai in 2017. He has over 19 years' teaching and research experience in various institutions, and is presently working as a Professor in the School of Computer Science and Engineering at Jain (Deemed to be University). He has published 20 SCI articles and more than 30 Scopus articles in peer reviewed journals. He has published seven Indian patents and received one Innovation Grand patent from Germany's Patent Office. He has actively participated as a Corresponding Editor for CRC Press, IGI Global Publications, and Benthom Science Publications and a reviewer in some of international journals; he is a member of IEEE, IAENG, and CSI societies. His main research interest are in the area of wireless sensor networks, Internet of Things, artificial intelligence, and machine learning.

Rajesh Kumar Dhanaraj is a distinguished Professor at Symbiosis International (Deemed University) in Pune, India. Before joining Symbiosis International University, he served as a Professor at the School of Computing Science & Engineering at Galgotias University in Greater Noida, India. His academic and research achievements have earned him a place among the top 2% of scientists globally, a recognition bestowed upon him by Elsevier and Stanford University. He earned his B.E. degree in Computer Science and Engineering from Anna University Chennai, India, in 2007. Subsequently, he obtained his M.Tech degree from Anna University Coimbatore, India, in 2010. His relentless pursuit of knowledge culminated in a Ph.D. in Computer Science from Anna University, Chennai, India, in 2017. He has authored and edited over 50 books on various cutting-edge technologies and holds 21 patents. Furthermore, he has contributed over 100 articles and papers to esteemed refereed journals and international conferences, in addition to providing chapters for several influential books. Professor Dhanaraj has shared his insights with the academic community by delivering numerous tech talks on disruptive technologies. He has forged meaningful partnerships with esteemed professors from top QS-ranked universities around the world, fostering a global network of academic excellence. His research interests encompass machine learning, cyber-physical systems, and wireless sensor networks. Professor Dhanaraj's expertise in these areas has led to numerous research talks on applied AI and cyber physical systems at various esteemed institutions. He has earned the distinction of being a senior member of the Institute of Electrical and Electronics Engineers (IEEE). He is also a member of the Computer Science Teacher Association (CSTA) and the International Association of Engineers (IAENG). Professor Dhanaraj's commitment to academic excellence extends to his role as an Associate Editor and Guest Editor for renowned journals, including *Elsevier Computers and Electrical Engineering*, *Human-Centric Computing and Information Sciences*, *Emerald–International Journal of Pervasive Computing and Communications*, and *Hindawi–Mobile Information Systems*. His expertise has earned him a position as an Expert Advisory Panel Member of Texas Instruments Inc., USA.

Yassine Maleh is an Associate Professor of Cybersecurity and IT Governance at Sultan Moulay Slimane University, Morocco. He is the founding chair of IEEE Consultant Network Morocco and the founding president of the African Research Center of Information Technology & Cybersecurity. He is a senior member of IEEE and a member of the International Association of Engineers IAENG and The Machine Intelligence Research Labs. Dr. Maleh has made contributions in the fields of information security and privacy, Internet of Things security, and wireless and constrained network security. His research interests include information security and privacy, the Internet of Things, network security, information systems, and IT governance. He has published over 200 papers (book chapters, international journals, and conferences/workshops), 30 edited books, and 6 authored books. He is the editor-in-chief of the *International Journal of Information Security and Privacy*, and the *International Journal of Smart Security Technologies* (*IJSST*). He serves as an associate editor for *IEEE Access* (2019 Impact Factor 4.098), the *International Journal of Digital Crime and Forensics* (*IJDCF*), and the *International Journal of Information Security and Privacy* (*IJISP*). He is a series editor of *Advances in Cybersecurity Management*, by CRC Press/Taylor & Francis. He was also a guest editor of a special issue on *Recent Advances on Cyber Security and Privacy for Cloud-of-Things* of the *International Journal of Digital Crime and Forensics* (*IJDCF*), Volume 10, Issue 3, July–September 2019. He has served and continues to serve on executive and technical program committees and as a reviewer of numerous international conferences and journals such as *Elsevier Ad Hoc Networks*, *IEEE Network Magazine, IEEE Sensor Journal, ICT Express*, and *Springer*

Cluster Computing. He received the Publons Top 1% reviewer award for the years 2018 and 2019.

Daniel Arockiam is currently working as Associate Professor in ASET–CSE at Amity University, Gwalior, Madhya Pradesh, India. He has 12 years of experience in academia. He has completed his B.E and M.E both at Anna University, Chennai. He earned his Ph.D. at Shri Venkateshwara University. He has published 50 papers in various reputed journals and conferences. He is currently doing research on neural networks, deep learning, AI and ML, and related areas. He has membership in various professional bodies.

List of contributors

Harshvi Adesara
Unitedworld Institute of Technology,
Karnavati University, Gandhinagar,
Gujarat, India

S. Annamalai
School of Computer Science and Engineering,
Jain (Deemed to be University), Bengaluru, Karnataka

S. Babu
Department of Computing Technologies, SRM Institute of Science and
Technology, Kattankulathur, Chennai, Tamilnadu, India

Aditya Bhardwaj
B. Tech-AIML, Unitedworld School of Computational Intelligence,
Karnavati University
Gujarat, India-382422

S. Bhuvaneswari
PG and Research Department of Computer Science and Applications,
Vivekanandha College of Arts and Sciences for Women (Autonomous),
Namakkal, Tamilnadu, India

Ananya Bulchandani
Unitedworld School of Computational Intelligence,
Karnavati University, Gujarat, India-382422

Tanya Chhabhadiya
Unitedworld School of Computational Intelligence,
Karnavati University, Gujarat-382422

Rajesh Kumar Dhanaraj
Symbiosis Institute of Computer Studies and Research (SICSR), Symbiosis International (Deemed University), Pune, India

A. Gayathiri
PG and Research Department of Computer Science and Applications, Vivekanandha College of Arts and Sciences for Women (Autonomous), Namakkal, Tamilnadu, India

S. Gopinathan
Department of Computing Technologies, SRM Institute of Science and Technology, Kattankulathur, Chennai, Tamilnadu, India

Manivel Kandasamy
Unitedworld Institute of Technology, Karnavati University, Gandhinagar, Gujarat, India

A. Suresh Kumar
School of Computer Science and Engineering Jain (Deemed to be University), Bengaluru, Karnataka, India

M. Kumaresan
School of Computer Science and Engineering, Jain (Deemed to be University), Bengaluru, Karnataka, India

Varun Malik
Chitkara University, Chandigarh, Punjab, India

Ruchi Mittal
Chitkara University, Chandigarh, Punjab, India

S. Mohanapriya
Head, Assistant Professor, Department Of Computer Applications, Sona College Of Arts & Science, Salem, Tamilnadu, India

Umika Patel
Unitedworld Institute of Technology, Karnavati University, Gandhinagar, Gujarat, India

Vishva Patel
Unitedworld Institute of Technology, Karnavati University, Gandhinagar, Gujarat, India

S. Sabitha
PG and Research Department of Computer Science and Applications, Vivekanandha College of Arts and Sciences for Women (Autonomous), Namakkal, Tamilnadu, India

M. Sathiya
PG and Research Department of Computer Science and Applications, Vivekanandha College of Arts and Sciences for Women (Autonomous), Namakkal, Tamilnadu, India

G. Sathya
PG and Research Department of Computer Science and Applications, Vivekanandha College of Arts and Sciences for Women (Autonomous), Namakkal, Tamilnadu, India

Ramesh Sekaran
School of Computer Science and Engineering, Jain (Deemed to be University), Bengaluru, Karnataka, India

Raju Shanmugam
Unitedworld Institute of Technology, Karnavati University, Gandhinagar, Gujarat, India

Devesh Pratap Singh
Department of Computer Science and Engineering, Graphic Era (Deemed to be University), Dehradun, Uttarakhand, India

Vikram Singh
Department of Computer Science and Engineering, Chaudhary Devi Lal University, Sirsa, Haryana, India

Palakshi Sinha
Unitedworld Institute of Technology, Karnavati University, Gandhinagar, Gujarat, India

P. Sumitra
PG and Research Department of Computer Science and Applications, Vivekanandha College of Arts and Sciences for Women (Autonomous), Namakkal, Tamilnadu, India

Siddhant Thapliyal
Department of Computer Science and Engineering, Graphic Era (Deemed to be University), Dehradun, Uttarakhand, India

Mohammad Wazid
Department of Computer Science and Engineering, Graphic Era (Deemed to be University), Dehradun, Uttarakhand, India

Preface

The world is experiencing a new revolutionary computing paradigm that has the potential to transform how people interact with mobile devices, physical places, personal computers, and stakeholders. This new emerging technology, known as pervasive computing (ubiquitous), envisions a society in which sensors, embedded processors, computers, and digital communications are all available as low-cost commodities. Users will have a convenient and comfortable environment thanks to ubiquitous computing, which combines computational and physical infrastructures into a single habitat. More than thousands of sensor devices will populate this environment, which provides novel functionality and specialised services that will increase interaction.

Mobile-friendly digital information and services are becoming more widely available. There is an increasing trend toward ubiquitous computing, which is described as processing, transmission, creation, processing, and storage of both visible and invisible information. The sensor objects are becoming smart, which are globally connected via a network, and interact with the users of the environment. Security and privacy are two of the most important needs for ubiquitous computing.

Artificial intelligence and machine learning has been developed as a real-world tool in quantifiable medical exercises to integrate analytics and predictive competencies. These learning algorithms necessitate a massive volume of data to learn and fortunately the globe is data-rich. Developed or developing countries would make a vital investment in the healthcare sector as it is considered the priority of any nation. Certainly, the artificial intelligence and machine learning field with this open science culture will redefine the association between researchers, scientists, and corporates and make an influence in medical science.

This book brings a broad discussion on the essential developments in ubiquitous computing considering them from economic, social, and technical perspectives. It clearly examines different sectors and application areas that benefit most from the potential of pervasive computing. It also brings new challenges and future directions of ubiquitous computing that require active complex solutions and pervasive management.

Chapter 1

Understanding the security, privacy, and surveillance conundrum in "digital India"

Vikram Singh[1], Varun Malik[2], and Ruchi Mittal[2]
[1]Department of Computer Science and Engineering, Chaudhary Devi Lal University, Sirsa, Haryana, India
[2]Chitkara University Institute of Engineering and Technology, Chitkara University, Chandigarh, Punjab, India

1.1 BACKGROUND

According to a *Statista* report, the cyberspace population reached more than half the world's human population with 4.66b active Internet users (59.5% of the human population) as of January 2021. Given these figures, cyberspace is the new place for social, political, and economic activities – private and public, or both [1]. With around 60% of the human population roaming the Internet actively, cyberspace inherits all the traits of homosapiens, including indulgence in criminal activities – called cybercrimes.

Experts attribute the quantum increase in cybercrimes in post-demonitisation[1] India (Table 1.1) to an increase in electronic transactions forced by the sudden withdrawal of banknotes of ₹500 and ₹1000 denominations from circulation on November 8, 2016 midnight. A whole lot of cyber-illiterate citizens had to resort to electronic modes of transactions, becoming vulnerable to cyber quacks [2]. News related to cybercrimes and related issues regarding the breach of privacy rights and an electronic surveillance regime, particularly the unauthorized, targeted, as well as mass surveillance, which was usually housed deeper inside a newspaper, hit the headlines in the media in mid-July 2021 when *The Wire* published an investigative story on the *Pegasus Project* based on the findings of a consortium of 17 media groups spread across the globe [3].

Securing cyberspace requires its ecosystem to be understood, which comprises (1) the Internet and telecommunications infrastructure as parameterised by tele-density in terms of telephone and mobile connections, and the number of Internet, e-commerce, and social media users in the country; (2) a regulatory framework for governing cyberspace operations, like e-commerce transactions, electronic communications, digital signature certificates, etc.; (3) ensuring the security of cybercitizens in performing their daily chores in cyberspace, developing a trust in the cyberinfrastructure of the nation, and placing an emergency response mechanism to deal with cyber-threats; (4) providing a vibrant, resilient, and robust cyberspace ecosystem for carrying out economic activities; (5) establishing a robust

DOI: 10.1201/9781003362685-1

Table 1.1 Cybercrimes in India [2]

Year	Cases registered	Arrests
2005	481	569
2006	453	565
2007	556	583
2008	464	373
2009	696	551
2010	1322	1193
2011	2213	1630
2012	3477	2071
2013	5893	3301
2014	9622	5752
2015	11592	8121
2016	12317	7990
2017	21796	11601
2018	27248	13569
2019	44546	15212

and tech-savvy cyber-legal environment to make the citizens feel secure, criminals feel apprehensive and fearful, and governments feel disinclined and honest in not surveilling its innocent countrymen; (6) well-prepared society and forces to defend its territories in any war-like situation in the fifth dimension, namely, cyberwarfare; and (7) most importantly, preparing its society and citizenry for living a smooth and happy life in cyberspace by raising digital literacy to levels where a good majority of the countrymen can work (for vocation), understand (the working), and create content (for others' consumption) in cyberspace [4].

A cyberspace regulatory framework includes acts promulgated by parliament and the statutes, rules, and guidelines notified by the government to regulate cyberspace. In India, cyberspace security is regulated by its Cyber Act of 2000 amended in 2008, and various rule sets, policies, and strategies notified thereunder. Subsection 1.2.1.3 discusses this aspect in detail. In a recent interpretation of the Indian Constitution, the apex Indian court has held the citizens' right to privacy a fundamental right albeit subject to certain conditions. Breach of privacy in cyberspace has become an easier task, is more frequent, and has severe consequences. Cyber surveillance of select citizens and large groups by private outfits and government agencies is another facet of cyberspace. In India, surveillance is governed by the Indian Telegraph Act, 1885u/s 5(2) and Rule 419(A), notified under this act.

This chapter covers these three closely related aspects of cyberspace, namely, cybersecurity, privacy, and surveillance in the Indian digital space. Section 1.2 and its subsection are devoted to security concerning Indian cyberspace. Privacy aspects of citizens' rights in cyberspace and relevant law has been discussed in Section 1.3 and the cyber-surveillance regime in India has been discussed in Section 1.4, wherein subsection 1.4.3 has covered various aspects of the *Pegasus Project*.

1.2 INDIAN CYBERSPACE SECURITY

Cyberspace is best understood as the virtual environment wherein people communicate with each other and consume certain services provided by the service-providing entities through a network of computers. The array of services available in cyberspace varies from education to entertainment, to research, to insurance, to banking, to health, to sale-purchase, to the stock market, and to government. The global nature and round-the-clock operations of the marketplace offered by cyberspace is being used by individuals, governments, telecommunications, enterprises of all size, *supervisory control, and data acquisition* (SCADA) systems, and military in an intricate manner wherein drawing boundaries between different components is irrelevant. In the years to come, cyberspace is bound to grow more complex and inclusive [5].

1.2.1 Cyberspace ecosystem in India

Cyberspace in India, as we look back today, started taking shape way back in 1976 when the National Informatics Centre (NIC) supported the government by providing Information Technology solutions. In 1987, the Indian cyberspace further expanded when the NICNET – a wide area network (WAN) was established by NIC with nodes at every district headquarter in the country. Through the 1980s and early 1990s, NIC was used to provide email and Internet services through the *very small aperture terminal* (VSAT) technology. The Department of Electronics with the financial support of the United Nations Development Programme started the *Education and Research NETwork* (ERNET) in 1986. For about a decade, ERNET provided Internet and domain services to University Libraries based on both TCP/IP and OSI protocol stacks. Later, the use and support of OSI were discontinued and all Internet traffic was shifted to TCP/IP. Around 1989–90, with the withdrawal of the US government's control over ARPANET, the Internet spread outside of US boundaries swiftly. Availability of cellular mobile phone technology in the last decade of the 20th Century coupled with the appearance of smartphones and the plummeting of Internet data rates in the second decade of the 21st Century has resulted in an exponential increase in the number of Internet users and volumes of Internet traffic in India. These developments have given rise to the creation of a huge Indian cyberspace with the largest number of social media users and a very big e-commerce marketplace [4].

This following subsection discusses the key components that comprise any cyberspace ecosystem.

1.2.1.1 Cyberspace infrastructure

The cyberspace infrastructure includes Internet penetration in society in terms of mobile subscribers, Internet users, social media users, e-commerce users; and bandwidth, quality, and cost-to-user of the Internet. Table 1.2 shows the tele-density in rural and urban India in terms of absolute numbers and percentage subscribers.

As of January 2021, the number of mobile subscriptions in India was 1.1b (110 crores), an increase of 23m (2.3 crores) over the January 2020 number (Table 1.3). The count of Internet users in January 2021 was recorded at 624m (62.4 crores) with an increase of 47m (4.7 crores) over the January 2020 figure. On similar lines, the number of social media users added during the same period is 78m (7.8 crores). This remarkable increase in Internet and social media users could mainly be attributed to the switch over of businesses, education, consultancy, and socialisation activities like alumni meetings, marriages, and condolence meetings, etc., to the online mode in the wake of movement restrictions during the COVID-19 pandemic [7].

1.2.1.2 Cyberspace regulatory framework

It includes government policies and a strategical framework for the governance of cyberspace. Government initiatives and approaches towards public

Table 1.2 Tele-penetration in India – February 28, 2021 [6]

Particulars	Wired	Wireless	Overall
Telephone subscribers in million (urban)	18.47	639.24	657.72
Telephone subscribers in million (rural)	1.71	528.47	530.18
Total subscription in million	20.19	1167.71	1187.90
Urban tele-density (percent)	3.93	136.03	139.96
Rural tele-density (percent)	0.19	59.28	59.48
Overall tele-density (percent)	1.48	85.78	87.26

Table 1.3 Pan-India Internet and social media users [6]

Particulars	January 2020	January 2021
Mobile subscriptions (million)	1087	1110
Internet users (million)	577	624
Social media users (million)	370	448

and private investment in sectors like secure, seamless, and omnipresent cyberinfrastructure, impartial and transparent spectrum allocation policies, opening cyberspace for upcoming technologies, a mechanism for managing fake news and illegal cyber-activities, signing bilateral and multilateral cooperation treaties with other countries. Section 1.2.2 discusses this thread in more detail.

1.2.1.3 Cybersecurity

A conducive cyberspace is characterized by the level of security of its cybercitizens, cyberinfrastructure, preemption, and swiftness of its cyber-emergency response system. Providing a secure ecosystem takes formulation of cybersecurity strategies, building a credible emergency response mechanism, providing cybersecurity training to the personnel responsible for operating, maintaining, and securing critical information infrastructures in the public and private sectors, namely, finance, telecommunications, transport, power, defense, and nuclear.

India has as many as 36 central organisations to safeguard cyberspace provincial mechanisms. Further, India was amongst the first few nation states to adopt a National Cyber Security Policy, yet little has changed on the digital security turf since its propounding in 2013. In light of more and more commonplace Indian citizens (read non-tech-savvy) using electronic services for financial transactions post-demonetisation in November 2016, the Ministry of Home Affairs, Govt. of India, in December 2018 has outlined the measures to strengthen the country's cybersecurity pastures. Sensing a little success of the 2013 policy and ascension of new challenges like cloud services, use of artificial intelligence in foiling the security measures, and general data protection requirements as per global standards, a National Cybersecurity Strategy 2021 (Draft) has been drafted and adopted to replace the 2013 Cyber Security Policy [5].

Indian cybersecurity comprises the following elements [5]:

- National Security Council of India
- National Information Board
- The Information Technology Act, 2000 amended in 2008
- National Cyber Security Policy, 2013
- National Cyber Security Strategy, 2020 (proposed draft)
- Computer Emergency Response Team – India (CERT-In)
- National Technical Research Organisation (NTRO)
- National Critical Information Infrastructure Protection Centre (NCIIPC)
- Indian Cyber Crime Coordination Centre (I4C)

In the Global Cybersecurity Index 2020 (Table 1.4), India is placed at 15th position with a score of 97.5 out of 100 and is ranked at position 10.

Table 1.4 Global Cybersecurity Index 2020 [8]

Sr. no.	Country	Rank	Score	Position
1	United States	1	100	1
2	United Kingdom	2	99.54	2
3	Saudi Arabia	2	99.54	3
4	Estonia	3	99.48	4
5	South Korea	4	98.52	5
6	Singapore	4	98.52	6
7	Spain	4	98.52	7
8	Russian Federation	5	98.06	8
9	United Arab Emirates	5	98.06	9
10	Malaysia	5	98.06	10
11	Lithuania	6	97.93	11
12	Japan	7	97.82	12
13	Canada	8	97.67	13
14	France	9	97.6	14
15	India	10	97.5	15

1.2.1.4 Cyber-economy

The cyber-ecosystem of a country also comprises the economic activities in the related sectors like design, development, manufacturing, export, and import of ICT artifacts and e-commerce infrastructure to name a few. Unsafe cyberspace hinders the growth of economic activities in general [9].

According to [10], the value of the Internet economy across India has increased from 125b in FY 2017 to 250b FY 2020 (in billion U.S. dollars). In 2015–16, as much as 5.6% of India's gross domestic product was contributed by the Internet economy. In absolute terms, in 2020, $537.4b was contributed by the Internet economy in the gross domestic product of India. This makes a 16% share of the Internet economy in the national GDP-2020. The report in [11] further states that half of the Internet economy share came from smartphone apps ($270.9b). Phone users working with mobile apps contributed 70% to mobile traffic [11].

1.2.1.5 Cybercrime and relevant laws

Like any walk of public life, cyberspace is also marred by the ecosystem of crime and criminals operating in cyberspace. For cyberspace to be a worthwhile space, it must have a matching and dynamic legal framework to deal with commercial transactions and crime committed in cyberspace. Bringing the cybercriminals to book is a nightmare for the bricks-and-mortar model of justice delivery [2, 12, 13].

Table 1.5 Global cybercrime trends [14]

2013 global cybercrimes		2016 global cybercrimes	
Country	Share	Country	Share
USA	39%	USA	23.96%
UK	8%	China	9.63%
Angola, China, Italy, Turkey, Ukraine	3% each	Brazil	5.84%
India, Bangladesh, Brazil, Israel, Holland	2% each	India	5.11%

1.2.1.6 Cyber-literacy

Cyber-literacy involves imparting cognitive and technical skills for effective use of Internet resources through cyberinfrastructure and computer and Internet education is shown in the Table 1.5. In 2021, India had a rank of 73 out of 120 countries for Internet literacy [15]. It is thus essential to focus on digital awareness and education. For a cyber-ecosystem to thrive it must relate itself to the masses. It requires that a common citizen is literate in cyberspace activities. Further, for self-sustaining cyberspace, a country also needs to invest in higher education and research funding in cyber-technologies. Mass cyber-awareness programmes and large-scale training programmes for service staff, law enforcement workforce, and judiciary are also required [16, 17]. The main components of cyber-literacy include the following:

Working in cyberspace: It requires basic technical knowledge for accessing cyberspace through tools like web browsers, email clients, social networking apps, citizen/customer service portals of governments and banks, e-commerce portals, etc. In India, around 30m (3 crores) people were trained under *Pradhan Mantri Gramin Digital Saksharta Abhiyan* – a GoI digital literacy programme over a period of 3 years (2018–20). Of this, 12.6m people were trained in basic digital literacy in the FY 2020–21 alone [18].

Understanding cyberspace: Understanding the modus operandi of cyberspace helps appreciate the appropriate online behaviour, net etiquettes, honey traps, and common dos and don'ts while surfing and working in various domains of cyberspace. It also helps an individual in understanding the terms of use of the Internet and services and verification of genuineness of the Internet content.

Creating cyberspace content: This comprises using tools to design, develop, and distribute/upload the text, hypertext, images, videos, RSS feeds, and blogs on the Internet. Creating and uploading the active Internet content transforms an Internet consumer into a participating netizen.

1.2.1.7 Cyberwarfare

For many, cyberwarfare is no longer restricted to fiction books and sci-fi movies. In the future, wars shall be fought in the fifth dimension namely cyberspace, apart from four traditional warfare dimensions – land, sea, air, and space. A sound and safe cyber-ecosystem requires vibrant cyber diplomacy and foreign policy and preemption capabilities in the cyberwarfare domain.

Further, it is implicit that all the components comprising the cyberspace ecosystem grow harmoniously thereby meaning that technological developments in cyberspace; awareness, education, and research in cyberspace technologies; economic and commercial activities in cyberspace; cyber laws and cybersecurity policies and infrastructure; etc. all keep pace with each other, lest skewed cyberspace may result.

1.2.2 Regulatory mechanisms in Indian cyberspace

Primarily, the Indian cyberspace is regulated by the Information Technology Act, 2000 with its latest amendments affected in 2008 – almost 13 years back at the time of writing of this chapter. Thirteen years is quite a long time when developments in the Information Technology field are considered. Quite a few technologies, like cloud computing and storage, social computing, big data analytics, and blockchain technologies, have sprung up in the meantime. New technologies have provided several new avenues and modi operandi to cybercriminals. Various players in cyberspace are resorting to highly sophisticated ways to snoop on targets for varied purposes [19, 20].

Although the IT Act 2000 and its subsequent amendment in 2008, attempt to bridge the gap of proper national cyber laws in India, it could not come up to the expectations of the dynamically changing turf of Indian cyberspace. With the world of cybercrime becoming murkier by the day, exponential growth of social media in the second decade of the 21st Century, the present provisions of Indian cyber law seem wanting in teeth and expanse [21]. Further, the false sense of anonymity in cyberspace lends quite a good number of unsuspecting and innocuous Internet users to the net of cybercriminals. Major punitive provisions provided in the ITA Act 2008 to administer and govern cyberspace are presented in Table 1.6.

In addition to the Information Technology Act, certain illegal actions committed in the Indian cyberspace also invite provisions of the Indian Penal Code (IPC), 1860 (see Table 1.7).

In relation to the corporate operations, techno-legal directives including cyber forensics and cybersecurity diligence of the Companies Act of 2013 also are applied in addition to IT Act. The Companies (Management and Administration) Rules, 2014 drafted and notified under the ambit of the Companies Act 2013 have prescribed the guidelines regarding companies' cybersecurity obligations, lest the onus lies upon company top-level management [23].

Table 1.6 Punitive provisions of the Information Technology Act [22]

Cyber offence/ cybercrime description	Relevant section	Nature of offence	Penalty/punishment/ compensation up to
Damage to computer and ommunication system	43	Civil offence	₹10m
Failure to protect data	43A	Civil offence	₹50m
Failure to furnish a document/report to authority	44A	Civil offence	₹0.15m/instance
Failure to file periodic return to authority	44B	Civil offence	₹5k/day
Failure to maintain account/books/ record	44C	Civil offence	₹10k/day
Contravention not specified in ITAA2008	45	Civil offence	₹25k
Tampering with computer source documents	65	Cognisable bailable	3 year jail term OR ₹0.2m fine OR both
Failure to comply with the orders of controller	68	Non-cognisable bailable	2 year jail term OR ₹0.1m fine OR both
Failure to assist the agency	69	Cognisable nonbailable	7 year jail term AND fine
Failure to block public access when so directed	69A	Cognisable nonbailable	7 year jail term AND fine
Failure to decrypt data for law enforcing agency	69B	Cognisable bailable	3 year jail term AND fine
Unauthorized access to protected system	70	Cognisable nonbailable	10 year jail term AND fine
Offensive or false messages	66A	Cognisable bailable	3 year jail term AND fine
Receiving stolen computer	66B	Cognisable bailable	3 year jail term OR ₹0.1m fine OR both
Identity theft	66C	Cognisable bailable	3 year jail term AND ₹0.1m fine
Cheating by personation	66D	Cognisable bailable	3 year jail term AND ₹0.1m fine
Violation of privacy	66E	Cognisable bailable	3 year jail term OR ₹0.2m fine OR both
Cyberterrorism	66F	Cognisable non-bailable	Life term

(Continued)

Table 1.6 (Cont.)

Cyber offence/ cybercrime description	Relevant section	Nature of offence	Penalty/punishment/ compensation up to
Obscenity in electronic format	67	Cognisable bailable non-bailable (after 1st conviction)	3 year jail term AND ₹0.5m fine (1st conviction) 5 year jail term AND ₹1m fine (after 1st conviction)
Pornography in electronic format	67A	Cognisable bailable Non-bailable (after 1st conviction)	5 year jail term AND ₹1m fine (1st conviction) 7 year jail term AND ₹1m fine (after 1st conviction)
Pedophilia in electronic format	67B	Cognisable bailable non-bailable (after 1st conviction)	5 year jail term AND ₹1m fine (1st conviction) 7 year jail term AND ₹1m fine (after 1st conviction)

Table 1.7 IPC sections invoked in cybercrimes

Offensive action in cyberspace	IPC section invoked
Fraud	464
False documentation	465
Forgery for preplanned cheating	468
Damage of reputation	469
Using a false document as genuine	471

1.3 PRIVACY IN "DIGITAL INDIA"

Information and communication technologies in cyberspace affect human privacy in very different ways that were previously not even possible. An individual's privacy may be encroached upon in more ways than one and digital devices have only added teeth and expanse to such incursions upon personal privacy [24]. Modern smartphone devices are all potential surveillance devices with several apps having user permissions to access phonebooks, microphones, cameras, location, images, audio, and videos files stored on the device [25, 26].

Further, with ever-increasing social and economic interactions taking place online, individuals share their personal information while using various Internet services and particularly so while surfing social media networks. The personal information so shared include name, email, birthday, vocation, hometown, workplace and home addresses, real-time locations, interests, status, relationships, photos, videos, etc. Millions of smartphone users of free mobile apps run at a risk of their personal information being sold and misused by the promoters of free apps. Most Android-based smartphone users will recall the popup windows similar to the one shown in Figure 1.1, prompted by the *Viber* app seeking access to photos, media, and files. The punch in this prompt is that the app can only be used by clicking the "ALLOW" command button, but in absence of Data Protection Laws/ Rules, there is no legal remedy in the case of misuse of personal information by these apps. In the situation elucidated above, the importance of data and privacy protection gains importance.

Strangely enough, in Table 1.8, India is counted amongst the countries having proper data protection legislation in place, whereas at the time of writing this section the first fortnight of February 2022, the Joint Parliamentary Committee deliberating upon the Indian Personal Data Protection Bill has submitted its report to the Speaker of Lok Sabha – the lower house of Indian Parliament, on December 16, 2021, and the both the houses of the Indian Parliament are yet to debate the bill. In this context, it may not be misplaced to mention that India's Data Protection Bill was drafted (and submitted to the Government in July 27, 2018) by a committee headed by Justice B.N. Srikrishna. In the given situation and until India has its own Data Protection Act, in Indian jurisdiction, any act of data breach and harvest cannot be branded as illegal; it can, at best, be considered immoral or unethical, but certainly, not illegal. Accordingly, Indian users are left to fend for themselves without any legal remedy to protect them from the loss or misuse of their personal data [4, 28].

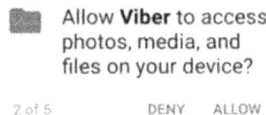

Allow **Viber** to access photos, media, and files on your device?

2 of 5   DENY   ALLOW

Figure 1.1   App seeks user permission.

Table 1.8 Worldwide status of Data Protection Laws (total 194 jurisdictions) [27]

Status of Data Protection Law	Number of countries	% of countries
Countries having Data Protection Legislation in place	128	66%
Countries having Draft Data Protection Legislation	19	10%
Countries having no Data Protection Law	37	19%
Countries providing no data	10	5%

1.3.1 Privacy law in India

In India, two cases, decided by its Supreme Court, have cast doubts on its citizens' right to privacy. In both the cases, namely, M.P. Sharma v. Satish Chandra of 1954 and Kharak Singh v. State of Uttar Pradesh of 1962, the apex court had held that privacy was not a fundamental right. These two decisions were pronounced by eight-judge and six-judge benches, respectively. But, in 2017, a nine-judge constitutional bench of the Supreme Court of India, in Puttaswamy v. Union of India, ruled that the right to privacy is a fundamental right, albeit, not absolute.

Thereafter, independent analysts have held informational privacy as one of the facets of privacy rights. Threats to the breach of information privacy may come from state as well as non-state actors. This warrants that Union Govt. place a personal information protection regime to safeguard the informational privacy of its citizens.

1.3.2 Privacy invasion in a digital age

The incursion on someone's privacy could happen in one of the following two ways [29]:

a. Intruding the personal privacy (breech of autonomy): This involves intrusion upon an individual's seclusion by way of snooping, recording, or listening to her/his private activities; or trespassing into an individual's private property. Concrete instances of this kind of intrusion include photographing someone in a try-room, call recording, hacking someone's account on a bank server or an email server.
b. Misuse of private information: This kind of breach involves disclosing an individual's private information, such as medical history or sexual orientation, without permission. Very often, there is an overlap in the two types of privacy invasion, and sometimes occurrence of one kind of privacy breach leads to another kind of breach as well.

Breach of personal information may have far-reaching repercussions on an individual's life. For instance, our personal information like customer names, Aadhaar number (India's national personal id), addresses, income and expenditure details, call details, Internet access details, travel details, purchase preferences, music and entertainment choices, credit score, and credit history, study loans, driving records, educational certificates, marital status, etc., are not stored on devices within our confines and control [29].

Another dimension of informational privacy is someone's medical and health-related information, wherein its breach might lead to life-shattering experiences, for example, disclosure of someone being HIV positive can ruin her/his social life and can even affect the professional life and workplace.

Personal information databases maintained by banks, NBFCs, Govt. agencies like UIDAI/Aadhaar pose problems in the realm of digital information privacy [5].

Search, watch, and seizure of someone's digital property and personal information is quite different from that of physical property. Digital information may be searched, copied, and (mis)used even without the owners being aware of it. Once the unauthorized control of digital resources is taken, existing digital information may be changed/manipulated and new incriminating material may be planted. Further, searching and removing a "banned" book or a copy of a periodical from circulation is difficult in comparison to tracking (and taking down) the digital content on the Internet [24, 30].

In the Bhima-Koregaon case of 2018, the digital forensic report by a US-based cyber-forensic lab "Arsenal Consulting" has claimed that the computer of one of the co-accused Ms. Rona Wilson was hacked, and as many as 22 computer files were planted days after the January 2018 violence. These files have been cited as evidence by Maharashtra police and the National Investigating Agency (NIA) of India [31].

Data mining and big data analytics can be used by state authorities to profile marginalised sections of society to draft welfare schemes or to snoop on them to preempt any impending uprising. China is using its "Police Cloud" to keep track of its religious and ethnic minorities. Social media companies have more access to users' personal information than the user herself/himself [32].

1.3.3 Privacy in digital India

Privacy concerns in the digital age could be attributed to state and non-state actors, or both. In the name of *all-important* national security, security agencies of the state profile their citizens based on their digital life. Profiling a whole lot of public segments through a *bricks-and-mortar* model may be impractical, whereas, digital profiling techniques are faster and less resource-intensive. Accordingly, states indulge in citizen profiling in the name of (and for the purpose of) national security. Of late, the capabilities of non-state actors to process and (mis)use has increased dramatically. For instance, Uber knows where we visit frequently. In case, we use Google online maps, then Google knows we stopped for a roadside tea or a breakfast [33].

Although, specific personal data and information protection law has not been promulgated in India as yet (July 31, 2021), sections 43, 66(E), and 72 specify the penalty and punishment for violation of privacy.

Section 43, although not directly addressing the privacy issue, provides for payments of compensation up to Rs. 10 million for downloading, copying, extracting, destroying, deleting, or altering any data or information that may cause a wrongful loss or wrongful gain to a person.

Section 66E addresses, albeit limited to images of the private area of a person, the issue of *punishment for violation of privacy* in the digital space. Anybody who intentionally, and without her/his consent, captures, publishes, or transmits the images of her/his private area is liable to be punished with a jail term of up to 3 years or a fine up to Rs. 0.2 million, or both.

Section 72 specifies the *"penalty for breach of confidentiality and privacy"*. A person who discloses to other person(s) "any electronic record, book, register, correspondence, information, document or other material without the consent of the person concerned" in a manner not otherwise allowed by the IT Act, shall invite punishment of a jail term up to 2 years, or a fine up to Rs. 0.1 million, or both.

Regarding the Indian Personal Data Protection Bill, the Joint Parliamentary Committee has yet to submit its report and there are bleak chances that the Bill will be tabled in the parliament any time before it meets for the winter session of 2021.

1.4 CYBER SURVEILLANCE IN INDIA

Beginning in the second decade of the 21st century, the government agencies in India have introduced a wide array of cyber-surveillance measures, marking a shift from targeted surveillance through traditional means to mass surveillance through digital means. Surveillance, for the purpose of the present discussion, may be classified in the following two ways: (i) government and private surveillance, and (ii) targeted and mass surveillance. In this section, governments' targeted and mass cyber-surveillance apparatus, in general, and the Indian Union, in particular, have been discussed [9].

Reasonable expectation of privacy to protect human dignity has been acknowledged and incorporated to varying extents in the constitutional frameworks of democracies across the globe. But, since the writing of constitutions and the digital revolution are not temporally synchronised, states' security agencies don't apply the same standard to cyberspace searches as are usually applied to the search of physical evidence. In the hyped national security environment all over the world, efforts of digital tuning of legislative frameworks are not paying due attention to digital privacy concerns [29].

In India, the vast powers of the Union and Provincial governments to snoop on its citizens stem from the Indian Telegraph Act, 1885, and the Information Technology Act, 2000. While the former law was promulgated in the colonial era itself, the latter too has reflections of preindependence laws. To add to the brawl are the state acts like MCOCA, permitting the state government to intercept communications.

1.4.1 Targeted and mass surveillance

Targeted surveillance is a form of scrutiny aimed at a specific individual or an organization of interest upon the existence of a prior suspicion. In most jurisdictions, including India, there is a concept of lawful interception under certain circumstances. Lawful targeted surveillance is allowed in Indian jurisdiction under the Telegraph Act, MCOCA[2], and UAPA[3].

Mass surveillance encompasses an entire or a major part of a population or an ethnic group to monitor their social, economic, political, or even religious interactions. This kind of surveillance is often carried out by agencies on the part of provincial and federal governments. United States' NSA's project PRISM belonged to this category. The legality of mass surveillance projects is decided by the law of the land, but as a rule of thumb, mass surveillance is an indicator of a totalitarian regime. Mass surveillance is often linked to compromised dignity and privacy of citizens, limited political and civil rights, as also to national security, terrorism, and social unrest. The remainder of this section discusses the various surveillance regimes in India. U/s 5 of the Indian Telegraph Act, 1885, the governments – Union as well as State – are empowered to take possession of any *telegraph*[4] or intercept any *telegraphic communication*[5] in the matters of public emergency, public safety, national security, sovereignty and integrity, friendly relations with foreign states, and to preempt an offence. The law, however, prerequires the Officials of the Ministry of Home Affairs to follow the procedure laid down in the law[6]. In India, although telephone tapping is considered a serious invasion of an individual's privacy and the right to privacy has been upheld as a fundamental right[7]. All in all, citizens' privacy right in India is not absolute but is subject to certain restrictions laid down by its apex court.

Under section 14 of MCOCA, interception of wire, electronic, or oral communications are permitted by the prior approval of competent authority, and the intercepted artifacts are admissible evidence in the court of law. Such permissions are allowed only to investigating officers of the rank of Superintendent of Police and further subject to rigorous prerequisites.

Also, UAPA has been alleged to interfere with a person's privacy in an arbitrary and unlawful manner. "Personal Knowledge" of the investigating officer without any written validation whatsoever from the judiciary suffices for searches, seizures, and arrests. UAPA also empowers the officers to intercept the accused person's communications without any independent prior approval or monitoring.

1.4.1.1 Lawful interception and monitoring (LIM)

Installation of *Lawful Interception* mechanism is a prerequisite for every licensed telecommunication network and Internet service provider. LIM is a secret mass cyber surveillance system deployed and operated by the Centre for Development of Telematics (C-DOT). This mechanism allows the GoI agencies to intercept voice and video calls, text and multimedia messages, Internet traffic, subscriber's particulars, call history, recharge and payment history, emails, browsing history, and other such user activities of Indian users. The LIM mechanism is deployed at the gateways of desired ISP(s) and its capabilities bypass the ISPs. Of late, as a licensing requirement, telcos in India have to deploy LIM on their own to enable the GoI to "lawfully intercept" the user traffic [34, 35].

LIM is carried out u/s 5(2) of the Indian Telegraph Act, 1885, read with rule 419(A) of Indian Telegraph (Amendment) Rules of 2007, notified under this Act. Further, on January 2, 2014, the Department of Telecommunication, GoI has issued Standard Operating Procedures (SOPs) for lawful interception of communications in the country [34, 35].

1.4.1.2 National Intelligence Grid (NATGRID)

The National Intelligence Grid was established in 2009 in the aftermath of the Mumbai terror attacks of 2008. NATGRID has access to real-time data collected by 21 different organizations, including banks, insurance companies, railways, immigration, telcos, and income tax department. NATGRID shares its information with 11 central agencies, which include the Research and Analysis Wing (RAW), Intelligence Bureau (IB), National Investigation Agency (NIA), Central Bureau of Investigation (CBI), and Narcotics Control Bureau (NCB), and Enforcement Directorate (ED) among others. NATGRID is a tool, and not an agency, that helps security agencies to find and locate terror suspects by using data sets pooled by 21 organisations mentioned above. NATGRID has been criticised on the two charges – usurping of privacy in the name of national security and efficacy in achieving its stated goal of preempting terror, for it does not share information with states' agencies [34, 35].

1.4.1.3 Centralised Monitoring System (CMS)

Centralised Monitoring System is a centralised mechanism of intercepting telecommunications bypassing the telecom operators to screen all kinds of traffic, including messages, voice calls, and internet usage. Software running at CMS server(s) is supposed to sniff the telecom signals to classify them into genuine communication and otherwise. Intricacies of traffic analysis algorithms have become very sophisticated with the incorporation of classification techniques empowered by data mining, big data analytics, machine learning, and deep learning. Figures 1.2 and 1.3 show monitoring mechanisms in the pre-CMS and post-CMS era, respectively [34, 35, 36].

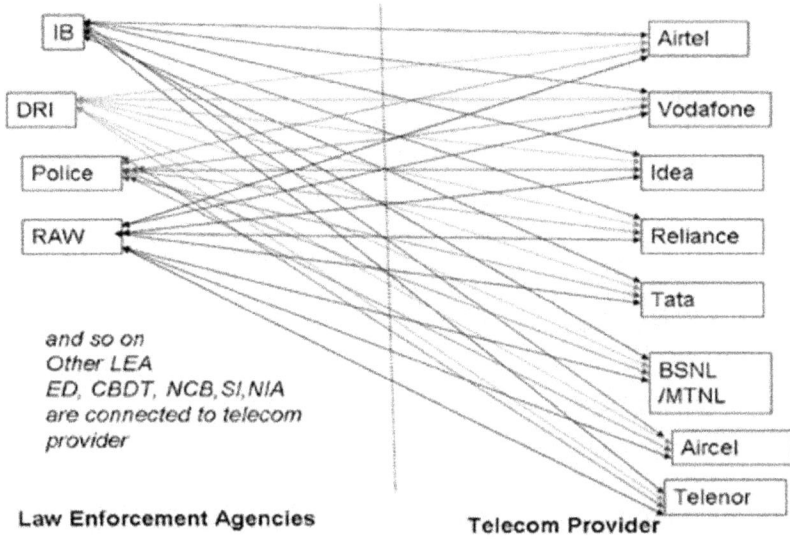

Figure 1.2   Indian surveillance regime in pre-CMS era [35].

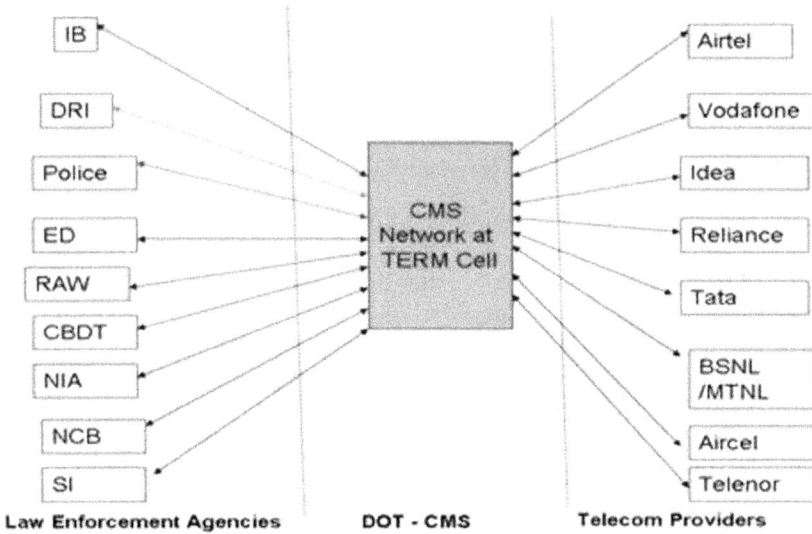

Figure 1.3   Indian surveillance regime in post-CMS era [35].

1.4.1.4 Network Traffic Analysis (NETRA)

NETRA is an Indian Government software system developed for inter-ception and analysis of internet traffic filtered using a predefined but dynamic logic. The system has been designed and developed by the Centre for Artificial Intelligence and Robotics (CAIR), a Defence Research and Development Organisation (DRDO) laboratory, and is used by the IB and RAW, India's internal and external security outfits, respectively. Modus operandi wise, NETRA is different from the CMS. NETRA watches every-thing the target users do online, whereas the CMS taps the communication network. NETRA can monitor the text contents of your emails, Facebook chats, and blogs using keyword-based filters [34, 35, 37].

1.4.1.5 National Cyber Coordination Centre (NCCC)

The National Cyber Coordination Centre (NCCC), an MHA entity, is a multiagency entity capable of performing assessing the cyber-threats in real-time. This centre generates reports whereupon law enforcement agencies can take proactive and preemptive actions [5], and mentions the multiplicity of central and state agencies that take care of Indian cyberspace. NCCC defends the country from cyberattacks by coordinating between multiple security agencies. According to industry standards, the centre may be classified as a mass surveillance system with its modus operandi exempted from the Right to Information Act – the transparency law of the land. According to Govt. sources, NCCC shall be the first line of defence against cyber-threats against the state's sensitive digital infrastructure.

1.4.1.6 National Technical Research Organization (NTRO)

NTRO has been established in the aftermath of the Kargil War – a limited armed conflict between the armed forces of India and Pakistan around mid-1999. As per the NTRO charter, the organisation was assigned to plan, design, set up, and operate new technical intelligence facilities as well as establish secure networks to connect various intelligence agencies and develop advanced cyber-intelligence techniques like cryptanalysis within India for cyber-defensive and cyber-offensive capabilities. The organization has a classified budgetary allocation and operates under NSA and PMO [34, 35].

1.4.2 The Pegasus Project

The world got to know of iOS exploitation by Pegasus[8] in August 2016 when Ahmed Mansoor, an Arab human rights activist received a text message having a hyperlink to secretive information about torture in UAE prisons. Investigation of the hyperlink by the Canada-based Citizen Lab confirmed it as social engineering spyware. The investigation attributed the spyware to the NSO Group and the spyware code contained entries relevant to Apple's iOS7 released in 2013.

In India, people heard the name of Pegasus in 2019 when *WhatsApp* notified that its communications might have been compromised by the Pegasus spyware during a 12-day time window stretching from April 29, 2019 to May 10, 2019. Another controversy involving Pegasus started on July 18, 2021 when *The Wire* broke a story on *The Pegasus Project* – a collaborative investigation involving more than 80 journalists belonging to 17 media groups [3] based in 10 countries spread over Europe, the Americas, Middle-East, and South Asia. The project has been coordinated by the *Forbidden Stories* of France. Cyber-forensics tech-support was lent by Amnesty International's Security Lab and the results were peer-reviewed by Canada-based *Citizen Labs* [3]

The data (about 50000 phone numbers) were accessed by the Forbidden Stories and Amnesty International, who then shared it with partner media groups for attributing the phone numbers to individuals for further investigation. Based on the analysis of these phone numbers the NSO potential clients were segregated into 11 countries, namely, Azerbaijan, Bahrain, Hungary, India, Kazakhstan, Mexico, Morocco, Rwanda, Saudi Arabia, Togo, and the UAE – in alphabetic order. However, just being on the list does not imply that the mobile number has been successfully targeted or even a hack has been attempted [3].

The Wire, the Indian partner in the project could attribute credible links of more than 300 (less than one-third of the total numbers with India code) India numbers to individuals who have used those phone numbers. *The Wire* team approached some 40–50 of about 300 identified persons and could manage to get the devices of only 21–22 of them for digital forensic analysis at Amnesty Security Labs. Of the forensically examined India-number phones, seven retained evidence(s) of actual infection by Pegasus, and three had evidence(s) of attempted infection [3].

While listening to a petition filed in the Pegasus row, the Supreme Court of India instituted a three-member special investigation team (SIT) to look into the matter. The SIT, vide a public notice, has requested the persons allegedly affected by the Pegasus spyware to submit their devices (iPhone) for capture of digital evidence, but, initially, only two devices were handed over to the SIT, and the time window in this regard had to be extended. Experts attribute the dismal response to the "public notice" to a highlighted phrase in the notice requiring "affected" person to mention "reasons as to why you believe your device may have been infected with Pegasus malware".

1.5 CONCLUSION

The Internet was conceived, designed, and evolved with security as one of the top concerns. It took off as a communication network and evolved as a data and resource-sharing platform. The prime object in those days was to establish a resilient and robust communication network that does not come down wholly in the eventuality of partial damage. It can easily and safely be theorised that the Internet is not secure by design. And, with the global cyberspace almost half the size of the global human population (and ever-increasing), cyberspace inherits all the traits of human society – crime inclusive. With India moving rapidly towards digitalisation in all walks of public life ranging from education to entertainment, from defence to governance, from commerce to banking to NBFCs, the cybercrime cases are spiraling high, in both absolute and relative terms. Furthermore, as the most common use of digital means is daily chores, the security of Indian cyberspace has taken centre stage of public discourse.

Further, with no specific cybersecurity legal framework to secure the Indian cyberspace, the Information Technology Act comes to the rescue of victims of cybercrimes and it proves dismally inadequate to address the issues arising in cybersecurity. The primary object of the Information Technology Act and subsequent in 2008, wherein digital signatures were introduced, was to pave the way for e-commerce and payment issues therein. Moreover, in the meantime, several new information technologies like the Internet of Things, cloud computing and storage, big data analytics, machine learning, and deep learning, etc., have graduated from the research to application arena.

Accordingly, to be on a par with technological developments and to bridge the gaps between technology adoption and the corresponding legal framework, and reassure the individuals and organisations operating in cyberspace, the Information Technology Act needs to be amended and a separate special cybersecurity law dealing with different facets of Indian cyberspace also needs to be promulgated.

With the IoT becoming mainline, Internet-enabled devices have become ubiquitous, in homes and at workplaces. Internet-enabled appliances and other devices are assigned IP addresses so that they could be operated and controlled remotely, for example, while driving home you can issue voice commands to your car to switch on the air conditioner(s) of your home so that you will find your home cool when you arrive, or you can open the main entrance to the house without alighting your car. An individual's privacy may be encroached upon in more ways than one. But, at the same time, these capabilities add to cyber-threats and possibly incursions in your at-home privacy. The IoT and digital devices have only added teeth and expanse to such incursions upon personal privacy. Above all, modern smartphone devices are all lethal surveillance devices with several apps having user permissions to access the content of the devices and I/O devices to receive and transmit the data using the smartphone.

With the focus shifted to government surveillance in wake of the *Pegasus Project* telephone data leaks, it is high time the government's projects with the potential to surveil its masses are debated and questioned. Pegasus leaks themselves need to be probed at an appropriate level and truth must prevail so that the citizenry is reassured about their right to privacy, the robustness of India cyberinfrastructure, and the security and safety of their digital assets including devices, data, and software.

Even though in Shreya Singhal v. Union of India case decided in March 2015, the Supreme Court of India has struck down the subsection 66A of the Information Technology Act, the government in July 2021 has submitted in the apex court that around one thousand first information reports (FIRs) have been registered all over India. Such kinds of aberrations detrimental to citizens' right to free speech must be dealt with sternly. The Indian civil society and the so-called fourth pillar of democracy – the media, must raise personal privacy and security issues vehemently so that privacy and security breaches by the governments and their agencies may be exposed and contained.

Notes

1   The Government of India, on November 8, 2016, announced the demonetisation of all banknotes of ₹500 and ₹1,000 denomination previously issued by its central bank. Further, the Government also introduced new banknotes of ₹20, ₹50, ₹100, ₹200, ₹500, and ₹2,000 denomination.

2   Maharashtra Control of Organised Crime Act, 1999 (MCOCA) was firstly promulgated as an ordinance and subsequently ratified by the Maharashtra state legislature. The Bill became the law with the assent of the President of India under Article 245 of Indian Constitution that applies to subjects in concurrent list.

3   Unlawful Activities (Prevention) Act, 1967 (UAPA) is aimed at preempting the happening of unlawful activities directed against the integrity and sovereignty of the Indian state.

4   The amended Telegraph Act defines the term "telegraph" as *"any appliance, instrument, material or apparatus used or capable of use for transmission or reception of signs, signals, writing, images and sounds or intelligence of any nature by wire, visual or other electro-magnetic emissions, radio waves or Hertzian waves, galvanic, electricor magnetic means."* In a digital age, taking possession of the telegraph translates into taking possession of communication and computing devices used to affect wired or wireless communications. This includes mobile phone, smartphones, tablets, notepads, laptops, personal computers, etc.

5   Telegraphic communications include the use of wired and wireless telegraphy, telephones, teletype, radio communications, and digital data communications. And, the instant law does not define the term "interception", therefore, the popular dictionary meaning of the word is adopted, i.e., an act of listening in and

recording communications, intended for another party, for the purpose of intelligence or counterintelligence.

6   In a 1996 decision in People's Union for Civil Liberties v. Union of India & Anr, the Supreme Court of India bench comprising Justices Kuldip Singh and Saghir Ahmad has stated that *"The first step under Section 5(2) of the Telegraph Act is the occurrence or happening of any public emergency or the existence of a public safety interest. Thereafter, the competent and concerned authority under section 5(2) of the Telegraph Act is authorized to pass an order of interception after recording its satisfaction that it is mandatory or expedient so to do in the interest of:*

  i. Sovereignty and integrity of India.
  ii. Security of the State.
  iii. Friendly relations with foreign states.
  iv. Public order.
  v. Preventing incitement or inducement to the commission of an offence.

In the instant case, the Supreme Court of India has termed the telephone tapping a serious invasion into individual's privacy, but allowed the lawful interception has been allowed under certain circumstances u/s 5 of the Indian Telegraph Act, 1885.

7   In the Puttaswamy v. Union of India case-decision delivered on August 24, 2017, a nine-judge constitutional bench of the Supreme Court of India has held the right to privacy as a fundamental right protected under Part III, Article 21 of the Indian Constitution.

8   Pegasus is a spyware developed by an Israeli cybersecurity firm NSO Group. As per the group's claims, the Pegasus is licensed only to Government organisations (intelligence agencies, law enforcement agencies, and militaries) vetted by Israeli Defence Ministry, to combat terror and crime. The software suit compromises and surveilles on the targeted devices running Windows, MacOS, Android, and iOS operating systems.

REFERENCES

1. J. Johnson, Worldwide Digital Population as of January 2021, Statista Report, April 1, 2021.
2. NCRB, *National Crimes Record Bureau*, Cybercrimes 2019.
3. S. Varadarajan, Revealed: How the Wire and its Partners Cracked the Pegasus Project and What it Means for India, *The Wire*, July 30, 2021.
4. A. Chhibbar, *Navigating the Indian Cyberspace Maze–Guide for Policymakers*, KW Publishers Pvt. Ltd., New Delhi, 2020.
5. V. Singh and V. Malik, Indian Cybersecurity Turf: A 2020 Position Paper, 2021.
6. TRAI, Highlights of Telecom Subscription Data as on 28th February, 2021, Telecom Regulatory Authority of India, Press Release No. 27 /2021.
7. R. Shankar, B.K Sarojini, H. Mehraj, A.S Kumar, R. Neware, A. Singh Bist, Impact of the Learning Rate and Batch Size on NOMA System using LSTM-Based Deep Neural Network. *The Journal of Defense Modeling and Simulation*, vol. 20, no. 2, 259–268, 2023. doi:10.1177/15485129211049782

8. *GCI-ITU, Global Cybersecurity Index 2020, English, ITU 2021*, Published in Geneva, Switzerland 2021, ISBN: 978-92-61-33921-0 (Electronic Version).

9. *State of Cyber Security and Surveillance in India–A Review of the Legal Landscape*, The Centre for Internet and Society. https://cis-india.org/internet- governance/blog/state-of- cyber-security-and-surveillance-in-india.pdf

10. S. Keelery, Internet Economy Value India: 2016–2020–Statista Report, October 16, 2020.

11. A. Mehbodniya, A. Suresh Kumar, K. P. Rane, K. K. Bhatia, and B. K. Singh, "Smartphone-Based mHealth and Internet of Things for Diabetes Control and Self-Management," *Journal of Healthcare Engineering*, vol. 2021, 2021.

12. S. Mehta, and V. Singh, Internet Usage and Cyber-Victimization in Indian Society, *Academic Discourse*, vol. 1, no. 2, pp. 124–131, 2012.

13. S. Mehta, and V. Singh, Combating the Cybercrime: Practices in Indian Society, *International Journal of Computing and Business Research*, vol. 3, no. 3, 2012.

14. IC3, Internet Crime Complaint Centre, Internet Crime Report 2020.

15. S. Keelery, Cybercrime in India–Statistics & Facts 2020–Statista Report, March 10, 2021.

16. M. Kandasamy, R. Shanmugam, and S. K. A, "Automatic Detection and Classification of Diabetic Retinopathy using Modified UNET," *2023 Third International Conference on Artificial Intelligence and Smart Energy (ICAIS), Coimbatore*, India, 2023, pp. 1468–1471. doi: 10.1109/ICAIS56108.2023.10073897

17. S. Mehta, and V. Singh, A Study of Awareness about Cyberlaws in the Indian Society, *International Journal of Computing and Business Research*, vol. 4, no. 1, 2013.

18. J. Shah, A Study of Awareness about Cyber Laws for Indian Youth, *International Journal of Trend in Scientific Research and Development*, vol. 1, no. 1, pp. 10–16, 2016.

19. A. Sureshkumar, R. Samson Ravindran, Swarm and Fuzzy Based Cooperative Caching Framework to Optimize Energy Consumption Over Multimedia Wireless Sensor Networks, *Wireless Personal Communications*, vol. 90, pp. 961–984, 2016. https://doi.org/10.1007/s11277-016-3274-0

20. M. Balamurugan, M. Kumaresan, V. Haripriya, S. Annamalai, and J. Bhuvana, "Secured Cloud Computing for Medical Database Monitoring Using Machine Learning Techniques", *Lecture Notes in Networks and Systems*, vol. 444, pp. 91–109, 2022.

21. A. Sarmah, R. Sarmah, and A.J. Baruah, A Brief Study on Cyber Crime and Cyber Laws of India, *International Research Journal of Engineering and Technology*, vol. 4, no. 6, pp. 1633–1641, 2017.

22. P.T. Shailesh, M. Nitin, and N.S. Shrikant, *A Review on Information Technology and Cyber Laws*, vol. 2, no. 15, pp. 10–16, 2015.

23. I.A. Sawaneh, F.K. Kamara, and A. Kamara, Cybersecurity: A Key Challenge to the Information Age in Sierra Leone, *Asian Journal of Interdisciplinary Research*, vol. 4, no. 1, pp. 35–46, 2021.

24. IT Act 2000 (ud.) *Information Technology Act 2000 Amended in 2008, Ministry of Electronics and Information Technology*, Government of India.

URL; www.meity.gov.in/writereaddata/files/it_amendment_act2008%20
%281%29_0.pdf

25. J. Kaur, and K.R. Ramkumar, The Recent Trends in Cyber Security: A Review, *Journal of King Saud University – Computer and Information Sciences*, vol. 34, no. 8, Part B, pp. 5766–5781. 2022. https://doi.org/10.1016/j.jks uci.2021.01.018

26. G. Rathinam, M. Balamurugan, V. Arulkumar, M. Kumaresan, S. Annamalai, and J. Bhuvana, Enhanced Security for Large-Scale 6G Cloud Computing: A Novel Approach to Identity based Encryption Key Generation, *Journal of Machine and Computing this Link is Disabled*, vol. 3, no. 2, pp. 80–91, 2023.

27. M. Ramanan, L. Singh, A. Suresh Kumar, A. Suresh, A. Sampathkumar, V. Jain, N. Bacanin, Secure Blockchain Enabled Cyber–Physical Health Systems Using Ensemble Convolution Neural Network Classification. *Computers and Electrical Engineering*, vol. 101, pp. 108058, 2022.

28. D. Lindsay, *The Right to Be Forgotten–European Data Protection Law, Emerging Challenges in Privacy Law: Comparative Perspectives*. Cambridge University Press, pp. 290–293, 2014.

29. S. Ardhapurkar, T. Srivastava, S. Sharma, V. Chaurasiya, and A. Vaish, Privacy and Data Protection in Cyberspace in Indian Environment, *International Journal of Engineering Science and Technology*, vol. 2, no. 5, pp. 942–951, 2010.

30. UNCTD (nd.), *Data Protection and Privacy Legislation Worldwide, The United Nations Conference on Trade and Development*.

31. R. Khan, *Cyber Privacy Issues in India*, 2013. DOI: http://dx.doi.org/ 10.2139/ssrn.2357266

32. W.T. DeVries, Protecting Privacy in the Digital Age, *Berkeley Technology Law Journal*, vol. 18, no. 1, pp. 283–311, 2003.

33. P.M. Schwartz, Privacy and Democracy in Cyberspace, *Vanderbilt Law Review*, vol. 52, pp. 1609–1701, 1999. URL: https://escholarship.org/uc/ item/2fq3v1mj

34. P.M. Schwartz, Internet Privacy and the State, Connecticut Law Review, vol. 815, 1999. URL: https://escholarship.org/uc/item/37x3z12g

35. N. Benwal, Bhima-Koregaon Case: Key Evidence Was Planted?, *The Free Press Journal*, Feb. 11, 2021.

36. Joint Civil Society Report. (2018), *Prepared by Network of Chinese Human Rights Defenders, submitted to The Committee on the Elimination of Racial Discrimination*, July 16, 2018.

37. *India's Surveillance State–Communications Surveillance in India*, Software Freedom Law Centre, URL: https://sflc.in/sites/default/files/wp-content/uplo ads/2014/09/SFLC- FINAL- SURVEILLANCE-REPORT.pdf

Chapter 2

A systemic review of RFID applications, broad casting areas, and public policy issues

M. Sathiya, G. Sathya, S. Sabitha, P. Sumitra, A. Gayathiri, and S. Bhuvaneswari
PG and Research Department of Computer Science and Applications, Vivekanandha College of Arts and Sciences for Women (Autonomous), Namakkal, Tamilnadu, India

2.1 INTRODUCTION

Radio frequency identification (RFID) is a wireless technique skilled at removing a single sign from electronic tags attached to objects automated with unmistakable classification while not in the line of vision. An electronic tag attached to an object called an RFID tag delivers information to a reader using radio waves with the goal of identifying and pursuing something. In a catastrophe situation, RFID is used to follow and trace innocent people.

RFID often captures information in real time and makes it available to emergency responders as an actual time saving. Through a database, crisis management teams, hospitals, and emergency workers can grant access to knowledge. RFID was the initial technology studied in the 1940s as a means of locating linked aircraft. Today, this mechanism is currently used in a variety of industries, including planning to produce, supplier chain management, transportation, agriculture, healthcare, and services, to name a few.

Large RFID applications are expected to offer major benefits in real estate, energy infrastructure, transportation safety, and healthcare across all industries and nations. RFID technology has progressed and altered over the last 50 years, becoming even more convenient and cost-effective for people in general, as well as a successful way of addressing structural and technological issues in a range of sectors. The requirement for standardisation, privacy, and systems is necessary across RFID domains, and other essential concerns are still being debated.

The RFID concept was first introduced during World War II. The size and location of an object was classifying as an object with the help of radars. Determining whether an aircraft was a friend or adversary was crucial. To address this difficulty a scientist developed a technique known as IFF (identification friend or foe). Every aircraft was equipped with a transponder that received signals from ground radar stations and sent back a signal as to whether the aircraft was a friend or foe.

This system formed the cornerstone of radio frequency identification. In 1945, the first RFID tag was invented by Leon Theremin, dubbed "the

DOI: 10.1201/9781003362685-2

Thing" an espionage device for Russia. It relayed incidental radio waves using audio data. It was given to the US in the form of a carved wooden plaque featuring the US national crest. For many years, Russia used the present as an agent while it was kept in the US embassy.

In 1948, Stockman demonstrated in his seminal study on "communication by means of reflected authority" that it is possible to modulate the signal by adjusting the proportion of reflected control by how unreliable the load of the tag antenna is. The term for it was antenna load modulation. RFID is a brand-new term for this type of wireless technology. As electronic and communication technology evolved, such as the use of transistors, integrated circuits, microprocessors, and communication networks, further advances were made in RFID systems. The design and installation of RFID systems was impacted by operational/business transformation and other applications.

Supply chain management expanded the number of applications, production, and verification of protocols and tracing systems. This tag is used to track the location of an object in a RFID system with more than one tag and reader by attaching an RFID tag to a piece of property that can be uniquely identified. Supply chain management for commercial operations can be managed efficiently and reliably with RFID technology. It can be used to ensure that the correct goods are offered in the correct area with no gaps or errors. Making relevant information available in real time can considerably improve administration and planning procedures.

The first game of the wearable RFID tag was invented by Zowie, and it was followed by Tagaboo, Ping Pong Plus, and Music Blocks.

RFID tracking of animal movement is an important application. It requires a significant amount of time and work to ensure that each individual cow in a herd of hundreds is properly fed. With RFID devices, however, this may be done automatically and economically. RFID can be used as an electronic key to open doors to restricted areas.

2.1.1 An outline of RFID technology's history

- RFID technology was created for defence purposes during World War II after Frederick Hertz eventually discovered its presence as frequency through an investigation in 1886.
- As more students and inventors became interested in the RFID system during the 1970s and 1980s, efforts to file patents advanced. Charles Walton and other researchers filed for a patent to use RFID.
- Several U.S. and European businesses began to produce RFID tags in the 1980s after realising the value together with developing RFID technology.
- As well as facilitating the usage of RFID technology, MIT students established an Auto-ID centre.

- The majority of students agreed that Waloutlet, who debuted an RFID-based material unique system in 2005, was in fact the company that first industrialised RFID technology.
- RFID technology has the potential to be a brand-new policy tool that will deliver high levels of openness, potency, and efficacy not only in different business sectors but also in the provision of public services.
- Table 2.1 gives a brief history of the development and use of RFID technology.

2.1.2 RFID system

An RFID system typically contains the following in the given Figure 2.1:

- RFID tag
- RFID reader
- Host system

Table 2.1 An overview of RFID technology

Date	Event
1886	From the town Hertz experiment, the idea of frequently using mirror waves from objects was born.
1930–1940	An IFF system was created by the analysis laboratories of the American navy (identify friend or foe).
1940–1950	During World War 2, the IFF system was the first application of RFID, and it was used to distinguish between friendly and terrorist aircraft.
1973	Former IBM scientist Charles Walton filed a copyright for an RFID enabled radio operated door locking system.
1980–1990	Numerous businesses in Europe and North America started making RFID tags.
2003	The Massachusetts Institute of Technology's Auto-ID Centre evolved into EPC International, a company whose purpose is to promote and utilize RFID technology.

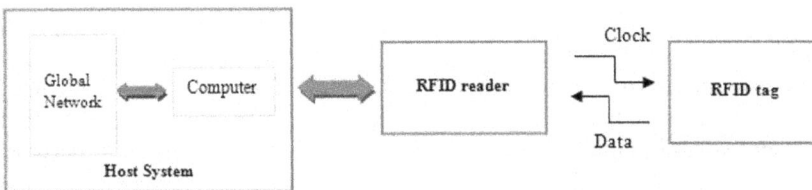

Figure 2.1   RFID system.

2.1.3 RFID tag

An item that has to be tracked or identified really does have an RFID tag affixed to it. An antenna and associated circuitry make up most tags. The onboard circuitry stores the identifying code as well as occasionally extra data such as the item's specifications or any handling instructions. RFID tags can be categorised depending on a variety of characteristics, such as operating frequencies, power sources, or the availability of a silicon chip. For RIFD tags, the frequency bands ultra-high frequency, high frequency, low frequency, and microwave makes all of them accessible. Inductive coupling is used; HF (13.56 MHz) and LF (120–150 KHz) RFID devices can communicate up to 1 m away. In comparison to LF and HF tags, UHF RFID tags typically operate between 866 and 868 MHz and 902 and 928 MHz. They have a greater scan range (up to 10 m) and a higher data rate. The frequencies 2.45 and 5.8 GHz, with a range of 1 to 2 m, are often used in the microware band for RFID devices. RFID labels are divided into passive labels, active labels, and semiactive labels based on the power source for functioning. A silicon chip and an antenna circuit are the main components of passive RFID tags. They lack an operational transmitter and an inbuilt power supply. The label is given inductive power by the electromagnetic signal passed by the RFID reader, enabling it to send its data.

2.1.4 RFID reader

An RFID reader typically consists of a transceiver and antenna. As part of the identifying procedure, it transmits interrogation signals to an RFID tag. According to its output power and radio frequency, the RFID reader connects with tags that are inside its interrogation zone. Data encoded in integrated circuit beacons is decoded by the drive and sent to the host computer for processing. Readers can be integrated into a hand-held device, such as a portable scanner, or they can be installed in fixed places around an enterprise.

2.1.5 The RFID system's operation

The majority of RFID systems use labels that are directly attached to goods to identify them. Depending on sorting and application, each label has an independent internal memory that can either be "read-only" or "re-written." RFID systems are being used in an increasing variety of applications, such as supply chain management, security and verification procedures, network monitoring, and handling, and so on. The question-answering beacons are launched by the reader's inquiry signal and high-frequency magnetic energy; the frequency of requests can exceed 50 times per second. As a result, most of the system's components, including the labels and the reader measurements, are now able to communicate with one another. As a result, enormous amounts of knowledge are produced. Therefore, giving industries access to the background information systems through square-measure victimisation filters manages this difficulty. In other words, code as initiated is used to manage this disadvantage. This code serves as a barrier between the RFID reader and the IT. Between the IT and, subsequently, the RFID reader, this code serves as a protective barrier.

2.1.6 RFID system subsystems

Two components that make up an RFID system are always present.

- The reader, which may be a read or write device depending on the blueprint and equipment used.
- It is positioned on the object to be recognised and contains a transponder.

The radio frequency identification system is composed of a variety of components that are joined in the way described in the section above. To carry out various actions with the RFID facilitates the ability to gain information. The RFID components are implemented by RFID solutions.

The RFID system is made up of the following five parts:

- Tag (attached to AN item, unique identification).
- Antenna (tag detector, produces magnetic field).
- Reader (manipulator of the receiver).
- System for communication (allows readers and RFID to navigate across IT infrastructure).
- User database, application, and interface for application software.

2.1.7 Types of RFID systems

RFID systems are classified into three basic categories:

- Low frequency (LF).
- High frequency (HF).
- Ultra-high frequency (UHF).
- Microwave RFID.
- **Low-frequency systems:** Although they have a regular frequency of 125 KHz, they can operate at any frequency between 30 and 500 KHz. LF RFID transmission ranges generally between a few inches and less than six feet.
- **High-frequency systems:** Even though they have a regular frequency of 125 KHz, they can function at any frequency from 30 to 500 KHz. The LF RFID transmission typically extends from a few inches to less than six feet.

- **UHF systems:** These systems can normally be read from a distance of 25 feet and also more and operate at frequencies ranging from 300 to 960 MHz, with 433 MHz being the most common.
- **Microwave systems:** This system can read from over 30 feet away and operate at 2.45 GHz.

For RFID application, frequency will differ significantly, and real distances may not always match assumptions. When the US announced that it would distribute e-passports containing RFID tag chips, it indicated that the RFID chips are only to be read from around 4 inches. However, the State Department quickly discovered that RFID readers could read data from RFID tags from far further away than 4 inches, in some cases as far as 33 feet. When larger read ranges are necessary, using more powerful tags can boost read ranges to 300 feet.

Why do we need an RFID system?

- **It boosts operating effectiveness.**

Most of the momentous advantages of RFID are that it needs less management and allows staff to focus on other jobs and further creative efforts. Additionally, understanding tags does not require a direct line of sight, so multiple technologies are able to read at an equal time. RFID readers can be programmed to immediately read tag information as needed.

- **It eliminates the possibility of human error.**

Manual work is always plagued with the risk of human error. RFID does not require any human intervention to read data. The reader may perform all of this automatically. RFID benefits significantly outweigh its limitations. RFID not only saves labour but also increases accuracy by avoiding errors resulting from human data entry and item for consumption restocking.

- **It lowers capital costs.**

It maintains tight control over stocks or assets, particularly costly company assets, such as test tools, transport covering cars, computer systems, and more. It is the simplest approach to keep fixed cost down. If any of these all of a sudden need to be replaced, it might be expensive. RFID technology makes it simple and within commercial means to track these assets.

- **It provides real-time data access.**

RFID features go beyond individual liberty. In harsh environments, RFID delivers robust track-and-trace. This system may simply collect this information and provide real-time product location data. Automatic real-time data collection can assist you in tracking large asset inventories, single commodities, or batches. RFID can also withstand difficult situations that typical barcode labels cannot, such as high humidity, large temperature swings, chemical and UV exposure, extremely high temperatures, and physical handling.

• **Provides insights to aid better decision making.**

RFID provides more insight and assists in making better decisions with real-time data to be fixed. RFID keeps you up to date at all times, which is essential when making planned and operational management decisions that could increase earnings.

2.2 REVIEW OF LITERATURE

Several North European businesses were involved in producing RFID information in the 1980s after seeing the value of developing RFID technology. To support the use and adoption of RFID technology, Massachusetts Institute of Technology University students have recently launched an Associate in Nursing Auto-ID centre. However, the majority of students claim that Walmarket Place, which introduced an RFID-based hardware differentiation system in 2005, was responsible for the majority of the development of RFID technology [1]. This technology presently uses their supply chain to track goods as well as clothing, technological items, and foods. RFID technology may be a cutting edge tool for policymakers, providing great transparency potency and efficacy not just inside manufacturing domains but also in government service deliverance.

The authors of [2] found that RFID has recently gained popularity due to its practicality and cost-effectiveness. Additionally, RFID is growing in acceptance due to its practicality and affordability. Additionally, RFID is now a technology that replaces barcodes. The widespread use of RFID is hampered, however, by a number of assaults and problems, including replay, spoofing, traceability, resynchronization, mismatch, and tag cloning. On RFID systems, there are two authentication mechanisms available. Another proxy-based authentication methodology is presented, along with proxy features in addition to how it aids in validation.

According to [3], RFID increases prepared remoteness from inches to a few feet or possibly hundreds of feet. When compared to traditional barcodes, RFID can operate at distances of up to hundreds of feet (with passive RFID tags) instead of just a few feet (active RFID tags). It is challenging to keep track of numerous RFID tags and find missing items linked to missing tags. A number of missing tag recognition procedures are designed using probabilistic methodologies and innovative run-time optimisation strategies.

The purpose of RFID technology in sequence security is that with clever, intelligent RFID tags placed in packaging, asset tracking is advanced to a new level. Readers may read the tags' data and update it automatically. An automatic technique for gathering product or transaction information is provided by RFID technology. Mechanisms of the RFID system are that a reader can read the data from tags (smart tags with built-in silicon chips that store up data), communications can be accumulated, and data can be found.

As of 2013, the authors of [4] explained and highlighted the main benefits of RFID technology. Access to a wide range of applications from different industries will be made possible via RFID. RFID technology offers a wealth of lucrative economic opportunities that can encourage various companies to embrace it, even though the investment cost is the main barrier to adoption. In the first part of the essay, the development of RFID technology is discussed, along with an explanation of how each component fits into the whole system. The end of the essay displays one of the many beneficial applications of RFID [5].

The authors of [6, 7] discussed the history of RFID technology development and how each component fits into the whole system in the first section of their article. The viability of using RFID technology is explored in the second section of the article, along with how it benefits from more productivity at lower prices. The final section of the essay illustrates just one possible use for RFID technology.

Aarti Singh and Manish Malhotra (2015) identified that the promise of cloud computing is that it may make transferring data and operations more affordable. Methods for securely outsourcing data with arbitrary computations currently in use either rely on the same piece of tamperproof hardware or on entirely homomorphism encryption, which has only recently been shown. Because fully homomorphism encryption is currently just of theoretical interest and is prohibitively expensive, a hardware-based solution is not commercially practical. We propose the architecture for offloading arbitrary computations to an untrustworthy commodity cloud while maintaining data safety.

The authors of [8, 9, 10, 11, 12] recognised that cloud computing has pushed Information Technology to new heights by providing the market condition of data storage and capacity and flexible scalable computing processing power to match fluctuating supply and demand while minimising capital expenditure. However, a successful benefit of cloud computing depends on properly managing security in cloud applications.

According to [13], RFID technologies have recently attracted considerable attention. The use of individual chip radios to tag actual objects, people, places, and things so they can communicate with computers has been the main driver of its expansion. As the backbone of the "Internet of Things" RFID technology is both lauded and insufferable, and in more fundamentalist circles, it is seen as the Mark. An RFID system can be analysed using two key dimensions. The technical infrastructure includes the actual data capturing system, which comprises of tags, readers, and a transmission medium. The term "logical infrastructure" refers to the complete identification (ID) system used to describe objects.

The authors of [14] identified that applications for RFID are still being developed and more are anticipated. There are other additional users for RFID tags that have already been developed, including security and commercial information. Another use for RFID tags is for people tracking. When RFID tags are sewed into uniforms, it's possible to monitor how long it takes an employee to finish a certain task. Numerous organisations are opposing the tracking of people via RFID because they are worried about how it would impact their social and private lives.

According to [15], a safe system is an absolute necessity, given the massive scale on which RFID determination function and the truth that it will penetrate every area of our lives. There are several stores that are experimenting with RFID, however it is still not extensively used, and there is an important body of literature on the topic. Maintaining the reliability of the tag and consequently, the produce, protecting the data stored on the tag, and protecting data that might be stored in a network database and is connected to the serial number on a tag are the three main issues with RFID and the requirement to protect confidential data.

2.2.1 Research design for a systematic review

We explored web databases and information gathered by industry specialists for RFID publications released between 2003 and 2015. By routinely reviewing published information, we categorise RFID applications and analyse the difficulties and challenges that RFID today faces. To start, the vast majority of research may be exposed by searching an electronic database. The homepage of the main library is the National Capital University, we may be able to access electronic databases such as World Net of Science (WWS), Google Scholar, Proquest Central, and Science Direct. We typically use the terms "RFID technology", "RFID government", "RFID application", and "RFID problem" when searching the literature [16, 17].

Typically, the majority of analyses can be located using this method. Our other preferred method of gathering information was to speak with specialists. To do so, we first prepared a list of specialists. A United Nations organisation focuses on public administration, science and technology, and information technology. Five specialists agreed to assist America and recommended certain study articles for their obvious logic and considerable content. We routinely choose relevant research papers based on the recommendations of experts. Overall, we employed the existing literature as our learning source, which we obtained via the two ways described above: searching an electronic database and interviewing specialists [18].

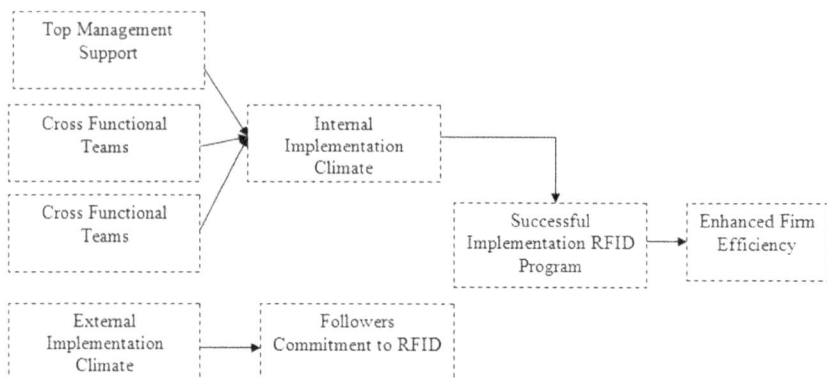

Figure 2.2   Analytical frame.

We chose the literature for a systematic review in line with the three processes shown on the flow chart. First, the e-database showed that there were 4260 various studies overall. Together, 185 analytical publications were found through peer-reviewed research and professional guidance. In total, 445 studies were choosen via the primary stage. We tended to eliminate four of the 121 applicants who satisfied the general qualifying requirements by showing the headings and concept. We typically only included RFID research that satisfied one of the following requirements:

- Studies having a private sector emphasis.
- Studies that exclude RFID required by the public sector.
- Studies that don't address the social science consequences.
- Studies that exclusively look at RFID technology from a purely technical and scientific standpoint.

Finally, we tend to mention the RFID difficulties that have been identified, as well as their implications for the public sector. Third, we tend to dismiss any of the 213 studies that are overly focused on the personal sector or RFID technology itself rather than the technology's social scientific applications and repercussions in Figure 2.2.

2.3 MAJOR AREAS OF RFID APPLICATION

2.3.1 Defence and security

To the best of our knowledge, as RFID technology has been developed over time, the requirement to make sure of national security has emerged is shown in Table 2.2. Nearly 60 years ago, ground forces produced an

Table 2.2 Defence and security of RFID

S. no.	Application	Case	Country
1.	Army and Navy	US Navy embedded RFID into cargo container	United States
2.		US Army track containers of material	United States
3.		US Army using RFID for tracking its army in Iraq	United States
4.	Airport port Security	Two RFID programmes to prevent terrorists attack	United States
5.		New York City Government project using RFID	United States, New York
6.	Prison management and child protection	Calpatria prison issued RFID embedded bracelet to its inmates	United States

RFID identify system to recognise allies and foes, and RFID technology has been being utilised to defend people. For example, the army and navy are deploying RFID technology on product containers to spot items. RFID is used by land forces to identify American weapons and containers. In addition, RFID systems are needed for airport security. The US President authorised the use of RFID-based identification systems at every landing field and port in America in order to defend the country from future terrorist strikes. In 2012, Taiwan's government chose to install a mixed RFID-based system, principally an electronic system, to improve security and electricity. Furthermore, technology of RFID is frequently employed efficiently in jail administration and kid protection. RFID labels are used to safeguard children in various nations [19, 20].

RFID security tag types

RFID security tags are classified into three types:

- Entry level
  - The entry-level tag is essentially an RFID chip with a barcode or other variable information printed on top.
  - This label has a self-adhesive backing that is applied to a separate swinging tag or packaging design.

- Inbuilt RFID swing tag
  - The inbuilt RFID swing tag can be hidden or visible.

- The chip is encased within the swing tag in the covert method, whereas the tag device is attached to the side of the swing tag in the overt option.
- Both alternatives can be customised with your branding or logo and other information you want to include.

- Integrated woven label

  - The integrated woven label RFID security tag resembles a typical woven label, however, it is made as a "pocket" that houses the device.
  - This woven pocket solution is typically of a generic nature and without any sort of marking, making it an excellent option to add the tag without detracting from the brand or product.

2.3.2 Applications in the environment

RFID technology will be used extensively in surrounding applications as shown in Table 2.3. Adoption of an RFID system-related certificate in throw-away management is the most significant RFID victimisation strategy in many nations across the world to verify economic and ecological waste management. The international organisation (EU) pioneered this sector ith

Table 2.3 RFID in environmental applications

Type	UHF	HF	LF
Range: (passive)	6 m	50 cm	Few cm
Benefits	Long range	Non artificial by water	Non artificial by water
	Standard	Non artificial by metals	Non artificial by metals
	Least rate	Multiple tag read	Frequency use without restriction
	Easy to save with low cost (5 cents)	Non affected by electrical noise	
Applications	Industry	Credit card, Access control card	Animal tracking
	Retail chain	Passports	Identification
Drawbacks	Absorbed by water	Range < 1m	Expansive
	Reflected by metals	Less efficient than LF	Noise
	Limited memory Interference with many applications		Low rate (70 ms)

its PAYT (pay-as-you-throw) programme. Pay-as-you-throw is an RFID throw-away evaluation mechanism with the intention that every person (and accompanying unit) has the total amount of garbage they dump monitored. The particular waste volume will be estimated as each unit and person integrates a waste hold into this integrated RFID label unit. It has been demonstrated throughout Europe that this incentive-based approach is a powerful policy instrument for lowering total waste and boosting activity [21].

An RFID tag is employed in waste management in monitoring nations, although the reason for acceptance in the United States differs from that in Europe. To deal with the constantly increasing number and varieties of garbage, India, the world's second most populated country, has implemented RFID technology. Similarly, in 2010, what softened China was the build-up of the planet and vast quantities of building garbage, which included a half-hour to forty-hour time frame for dealing with all urban waste. Shanghai was chosen for a waste management system based mostly on a related victimisation diploma [22].

The regime controlled the quantity of rubbish transported by all waste vehicles, which were equipped with RFID tags. Another great example of how it pertains to the environment is South Korea. A U Street tree in the government of South Korea maintains the precise location and position of street trees. RFID tags were applied to the street trees to find the location of tree in cities. An RFID tag was applied to each tree and entered in a net data system, allowing trees to be properly maintained. This network-based data system can manage remote information with the matching degree of interactive system most of the time [23].

2.3.3 Transportations

The RFID transponder is embedded into the RFID sticker to study the range of different elements, such as reader type, RFID frequency, and interaction with the neighbouring atmosphere or from other RFID tags and readers. The power source has the RFID tags with a much larger reading range.

A different conventional industry for RFID technology is public transportation. The RFID tag reads and scans the automobile, and the RFID reader scans and reads the information. The motive force can pay debits in accordance with the value recommended by the electronic reader. In the United States, the use of electronic tolls is seen as a cost-effective and expedient technique of avoiding long queues at toll booths. RFID-based toll grouping is increasingly being used in criminal cases since it allows prosecutors to pinpoint the exact position of a defendant's vehicle [24].

The Korean government has built the "Hypass" credit or debit card electronic toll collection system, which is primarily for the collection of transit on authorised routes. Drivers will avoid the tollbooth if they have an RFID tag installed on their vehicle because the RFID reader scans the data rapidly and completes the payment operation in around five seconds. Since Transport Canada implemented a similar mix of transit fees in 1997, the Octopus Card has gained international recognition for its convenience. The "Hypass" credit card electronic toll collection system is specifically developed for accumulating transit tolls in certain circumstances [25].

Furthermore, radio frequency identification is used as a serious technology in adopting countries to boost the potency and openness of public installations. The Mexican government, for example, is conducting "Creating Traffic Data in Mexico: Using RFID to Prevent Vandalism" and one of the goals of this one-of-a-kind article is to construct a transport data classification structure in order to collect a large amount of refined knowledge necessary for legislative decisions. In East Pakistan, the BRTA – Bangladesh Road Transport Authority – was established in 2003. The technology is mostly used as network administration with management, similar to the situation in Mexico. RFID technology was also used to collect train tolls in Bharat, where trains are the biggest extensively used means of common transportation.

2.3.4 Welfare and universal healthcare

RFID makes it possible for hospitals to better manage their equipment and save money in the public health sector. RFID tags have already been used in the pharmaceutical industry by federal government agencies. Since American hospitals provide nearly four different medications each day, medication errors are inevitable. Public hospitals in Taiwan have actively embraced RFID innovation with strong government support. Engineers inside Asia are creating an RFID identification system for those who are visually impaired, despite the fact that it has not yet been commercialised, with the assistance of the Pakistani government [26].

2.3.5 Agriculture and livestock

When it comes to managing agriculture, placental mammals, and ensuring the safety of food, RFID technology is frequently an effective instrument. The fact that disease tracing is frequently accomplished using cutting-edge technology like RFID is another important benefit of the current system. As a foundational framework for risk management in agriculture, researchers have created the Navigation System for Acceptable Chemical Use with government assistance. The European Union (EU) was the first to implement RFID technology in agriculture in a number of nations, including South Korea, Japan, Australia and quickly followed suit. The Australian government was the most committed to introducing RFID among those nations.

In Australia, for example, all newborn placental mammals have RFID tags implanted in them. The National Placental Mammal Identification System enables farmers to spot each animal and its health state. RFID technology has also been used in agriculture in Japan, including food safety and agricultural risk management due to over-use of pesticides. The Government of Japan has set itself the goal of building a food traceability system by 2010 under the e-Japan 2008 initiative. Another example of implementing a mandatory RFID is finding the system in placental animal management.

The Agriculture Department is focusing on RFID tagging to make tracking disease patterns easier. The NAIS was developed in 2002 with the establishment of the United States Institute for Animal Agriculture (NIAA). The United States government achieved this goal through this research because "the sooner animal health inspectors find ill and exposed animals and premises, the sooner they can manage the infection and restrict its spread" as stated inside the programme's purpose.

2.3.6 Public privacy in RFID diffusion

Applications of RFID raise complex policy and governance issues. They constantly deal with public issues for the technical gap in the ambiguity around the expected profit and expenses of rapid and widespread RFID adoption. Greek god RFID applications are raising concerns about transparency, digital identity, and power distribution. We routinely discuss governance concerns, such as corruption, privacy hazards, and digital monopolies and accomplishment.

2.3.7 Technologies concerns

Currently, technology is insufficient to solve all of the situations that RFID is seeking to address via a number of operational methods. RFID technology problems gradually emerge with technology utilisation due to their remaining restricted market. RFID technology, for example, lacks a standardised uniform frequency. Because there are no globally defined frequencies for RFID operations, allowable scanner/reader strengths vary by location. The frequencies from the EU and, as a result, the United States continue to diverge dramatically.

We became aware of the scarcity of storage space as well. In the RFID-based rubbish management widely employed, the technical constraint is a lack of storage capacity. Finally, from a toddler-watching incident that swept throughout Japan, it is concluded that knowing a child's specific location cannot assure their safety, despite the fact that RFID tags commonly give that intended illusion. Vining provided another warning about a specialised market risk. According to his research on port security in the United States, it is conceivable to steal a product at a port without breaking the RFID tag since the equipment may be trained and the contents taken out.

The RFID tag does not need to sustain any harm during this operation. During this process, there is no need for the RFID tag to suffer any damage. The US government implemented privacy protection in response to continuous requests from many groups.

2.3.8 Cost-benefit effectiveness is uncertain

RFID proponents claim the usability and potency of RFID technology will be ensured at a very low cost. However, there is compelling evidence that RFID will incur unforeseen costs. RFID has significant additional costs in comparison to barcodes, which have been frequently used to identify things before RFID technology was invented. Obtaining RFID hardware, tags, and devices is inadequate to successfully drive the system. The RFID system wants many lists such as "abundant mission, the luxury of authority, circular method mechanism, project manager, programmers" to assure a better grade of service.

2.3.9 Corruption and troubling transparency

RFID technology was supposed to improve transparency and reduce fraud. RFID technology, on the other hand, cannot provide greater transparency than planned. RFID tags may be easily duplicated and manipulated, and tag fraud will occur at each stage of RFID installation. Several examples demonstrate how RFID technology is being misused. Regarding the danger of tag corruption, RFID reader tampering has been identified, for example, where an RFID-based waste management system was actively enforced, a few people dumped their domestic garbage at their workplaces to avoid having it properly assessed by the RFID system.

The current protection model focuses mostly on the danger of tag fraud, but reader fraud will significantly impair users' privacy. One of the tagging corruption incidences that we discovered, according to Jules, was that in a firm, every single person who worked for an unscrupulous shop affixed a cloned tag to a bogus prescription. Avoine emphasised the prospect of reader contamination and claimed that web-based databases may be attacked directly. There are unethical methods of evading RFID detection technology. Some people in the EU have been disposing of garbage from their homes at work, where an RFID-based waste management system is being consistently applied.

2.3.10 Privacy issues

The main aim of the RFID technology is to secure the information enough to guarantee privacy is shown in Figure 2.3. The most critical issue that RFID

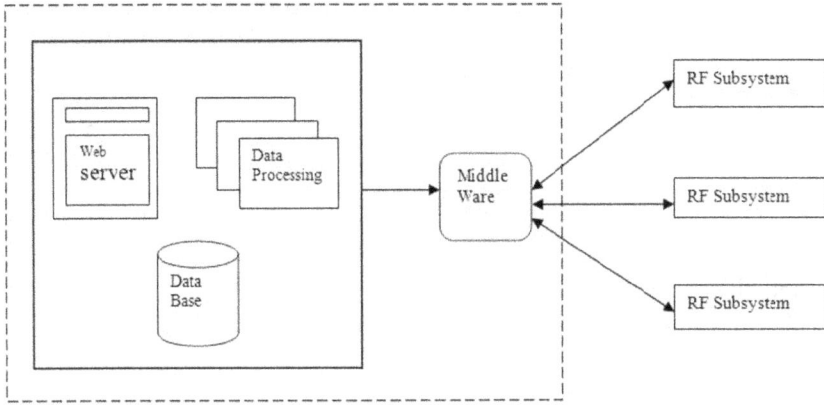

Figure 2.3   Privacy issues.

users must overcome is privacy. RFID-embedded chips typically include sensitive personal information, which can be compromised and can severely jeopardise someone's privacy. Engineers have developed encryption to keep personal data from escaping, yet there is still criticism. Various forms of privacy risks arise as a result of RFID technology's existing weak password protection capabilities.

As we mentioned before in the section on technological difficulties, present RFID technology is not designed to provide enough isolation. The technology itself has a number of faults, while some are astutely adequate to seek out specialised uses that would weaken the RFID security mechanism. RFID, by itself, causes a significant privacy concern, as well as other hidden expenses. Nonetheless, despite the threat of privacy breaches, a considerable number of stakeholders and students embrace the impending benefits of RFID. They claim that the usage of RFID chips for identification and tracking is entirely responsible for their successful implementation.

Mobile nursing care systems use RFID tags to improve the staff's ability to monitor patients' vital signs in a variety of settings and in various hospitals. In a mobile nursing setting, the Mobile Intelligent Medical System offers essential sign monitoring, mobile nursing applications, alarming services, and rule-based clinical decision support. To lessen the danger of severe harm brought on by the tardy delivery of medical care, the system constantly monitors critically unwell patients. Additionally, we think the system can be connected with other hospital information systems and expanded to the majority of medical areas. The MIMS will provide the healthcare sector with better and safer medical services as more medical facilities are connected to the system. The use of RFID technology having definite advantages, such as saving money, time, and resources while eliminating theft.

2.4 CONCLUSION

We talked about how airports and the military specifically manage RFID technologies to maintain security in the context of defence and security. Additionally, we discovered that RFID-identifying technologies like e-ID and e-passport. From wealthy to developing nations, RFID devices are frequently employed for managing rubbish and street trees. RFID-based cards have shown to be quite effective for the delivery of healthcare and welfare services. Monitoring fake medications is currently done with RFID. RFID has been used to track infections and provide services for the disabled. However, despite the potential advantages of RFID usage, certain unanticipated issues appear. RFID may still have technical flaws, particularly with regard to the security of cryptographic methods, global occurrence standards, and storage power. In some applications, RFID technology still lacks capability. Corruption of tags and readers can compromise security and transparency. The most major obstacles that RFID now faces are privacy issues. Even the most advanced technology, such as RFID, does not provide all of the answers for ensuring efficacy, efficiency, ease, and transparency. Instead, the deployment of an RFID system itself presents unanticipated problems. It should be emphasised that independent governance and hope continue towards it playing an important part in the innovation and resolution of policy problems created by the fast spread of RFID.

REFERENCES

1. Arefin AS Hossain MF, Sohel MK. "Designing and implementing RFID technology" in *Bangladesh*. *Dhaka*. Bangladesh: National Conference on Communication and Information Security; 2019.
2. Hande Bridelall R. "Novel RFID technologies: Energy harvesting for self-powered autonomous RFID" in *RFID Systems: Research Trends and Challenges*. Chichester, Wiley, pp. 473–495, July 2010. doi:10.1002/9780470665251.ch18.
3. Konsynski Smith H.A. "Developments in practice x: RFID-an internet for physical objects". *Communications of the Association for Information Systems*, 12, Article 19, pp. 301–311, 2013.
4. Brown D. "Part II: Applications" in *RFID Implementation*. New York, NY, USA, McGraw-Hill; 2017, pp. 133–222.
5. Naveen Kumar H.N, Suresh Kumar A, Guru Prasad M.S, Mohd Asif Shah, "Automatic facial b expression recognition combining texture and shape features from prominent facial regions". *IET Image Processing*, 2022. https://doi.org/10.1049/ipr2.12700
6. Ramanan M, Singh L, Suresh Kumar A, Suresh A, Sampathkumar A, Jain V, Bacanin N. Secure blockchain enabled Cyber–Physical health systems using ensemble convolution neural network classification. *Computers and Electrical Engineering*, 101, pp. 108058, 2022.

7. Jones E.C, Chung C.A. "Part 2: Integrating RFID into logistics". *A Practical Introduction*. New York, NY, USA: CRC Press; 2018.

8. Shankar R, Sarojini B.K, Mehraj H, Kumar A.S, Neware R, Singh Bist A. Impact of the learning rate and batch size on NOMA system using LSTM-based deep neural network. The *Journal of Defense Modeling and Simulation*, 20, no. 2, pp. 259–268, 2023. doi:10.1177/15485129211049782

9. Weigelt K, Hambsch M, Karacs G, Hubler A.C, Zillger T. "Labeling the world: Tagging mass products with printing processes". *IEEE Pervasive Computing*, 9, no. 2, 2019.

10. Tom K, Thomas L, Cheng S, Pecht M. "A wireless sensor system for prognostics and health management". *IEEE Sensors Journal*, 10, no. 4, 2019.

11. Kuo CH, Chen HG. "The critical issues about deploying RFID in healthcare industry by service perpective". In: *Hawaii International Conference on System Sciences, Proceedings of the 41st Annual*. Waikoloa, HI: IEEE; 2018.

12. Kumaresan M, Basha, M.J, Manikandan P, Annamalai S, Sekaran R, Kumar A. S. "Stock price prediction model using LSTM: A comparative study". *2023 3rd Asian Conference on Innovation in Technology (ASIANCON)*, Ravet IN, India, 2023, pp. 1–5. doi: 10.1109/ASIANCON58793.2023.10270708

13. Reindl L.M, Plessky V.P. "Review on SAW RFID tags". *IEEE Transactions on Ultrasonics, Ferroelectrics, and Frequency Control*, 57, no. 3, pp. 654–668, 2018.

14. Quaddus M, Hossain MA. "An adoption-diffusion model for RFID applications in Bangladesh". *International Conference on Computer and Information Technology (ICCIT)*, pp. 127–132, 2009.

15. Arumugam, Suresh Kumar, Amin Salih Mohammed, Kalpana Nagarajan, Kanagachidambaresan Ramasubramanian, S. B. Goyal, et al., "A novel energy efficient threshold based algorithm for wireless body sensor network". *Energies*, 15, no. 16, p. 6095, 2002.

16. Quaddus M, Hossain MA "Developing and validating a hierarchical model of external responsiveness: A study on RFID technology". *Information Systems Frontiers*, 17, pp. 109–125, 2016.

17. Hwang MS, Wei CH, Lee CY. "Privacy and security requirements for RFID applications". *Journal of Computers*, 20, no. 3, October 2009.

18. Jensen A, Cazier J, Dave D. "The impact of government trust perception on privacy risk perceptions and consumer acceptance of residual RFID technologies". *AMCIS 2007 Proceedings*, 146, pp. 1–8, 2017.

19. Kovavisaruch L, Suntharasaj P. "Converging technology in society: opportunity for radio frequency identification (RFID) in Thailand's transportation system". In: *Management of Engineering and Technology*, Portland International Center for. Portland, OR: IEEE, 2017.

20. Suresh Kumar A, Jerald Nirmal Kumar S. Subhash Chandra Gupta, Anurag Shrivastava, Keshav Kumar, Rituraj Jain, et al., "IoT communication for Grid-Tie matrix converter with power factor control using the Adaptive Fuzzy Sliding (AFS) method". *Scientific Programming*, 03, pp. 1–11, 2022.

21. Lee Y, Kuo F, Tang C.Y "*The Development of RFID in Healthcare in Taiwan*" Bejing: ICEB, 2014.

22. Chen Z.N, Qing X, Goh C.K. "Impedance characterization of RFID tag antennas and application in tag co-design". *IEEE T-MTT*, vol. 57, no. 5, pp. 1268–1274, May 2009.
23. Wyld D.C. "Delta airlines Tags Baggage with RFID". *Re:ID Magazine*. 2015.
24. Wyld D.C. "*RFID: The Right Frequency for Government*. Washington DC: IBM Center for the Business of government. IBM Center, 2005.
25. Wyld D.C. "Death sticks and taxes: RFID tagging of cigarettes". *Int J Retail Distribution Manag.*, vol. 36, no. 7, pp. 571–582, 2018.
26. Zhang R. A transportation security system applying RFID and GPS. *J Ind Engr Manag.*, vol. 6, no. 1, pp. 163–174, Special Issue: LISS 2012, 2013.

Chapter 3

A security and privacy approach for RFID-based wireless networks in ubiquitous computing

S. Gopinathan and S. Babu
Department of Computing Technologies, SRM Institute of Science and Technology,
Kattankulathur, Chennai, Tamilnadu, India

3.1 INTRODUCTION

Due to increased operational costs, manufacturers are looking for innovative methods to cut expenses and improve their operations. Radio frequency identification (RFID), a method for object recognition that transmits data without physical contact, could be a choice for mass production automation projects. The purpose of these wireless networks is to act as "smart monitoring systems" that offer thorough monitoring and tracking capabilities from manufacture to the supply chain's ultimate client phases. RFID uses wireless sensors, wireless technology, and communication network technologies to monitor and trace goods and material flow in manufacturing processes.

An organisation's part manufacturer can become more efficient and reliable by utilising RFID technology. RFID systems help reduce product and equipment shortages by gathering actual data, providing customers with dependable, on-time distribution or service, monitoring part inventories, and providing field maintenance histories. Security, quality assurance, production execution, and asset management are just a few of the production issues that RFID can handle.

RFID offers a simple approach to instantly gather data on a product, place, time, or transaction without the risk of human error. RFID technology is widely employed in a variety of fields, including healthcare, law enforcement, baggage handling, and toll roads, among others.

3.1.1 The elements of an RFID system

An RFID system can recognise items (tags) and carry out a variety of activities on them thanks to a variety of components [1]. By integrating RFID components, an RFID solution can be implemented. These three elements make up the RFID system.

- Tag.
- Reader.
- Backend database server.

DOI: 10.1201/9781003362685-3

3.1.2 Tags

Each object in an RFID system has a distinct ID, which is often entered into tags. A tag tool has microchips inside to keep an object's unique identification number. A silicon chip that is implanted with a microchip is a typical integrated circuit. The ID that is saved could be permanent or changeable relying on the microchip's read and write characteristics; tags can therefore be read-only or read-write data, depending on their classification. In essence, tags can be divided into three groups: active, semi-active, and passive. Semi-active tags combine active and passive characteristics, whereas active tags have either a full or a partial battery. How successfully RFID tags do specific tasks depends on their characteristics, including their frequencies, bandwidth, storage, information, and privacy.

3.1.3 Risks to privacy and security

Privacy and security issues with RFID tags may affect both individuals and organisations. Unprotected tags may be vulnerable to numerous dangers, such as traffic analysis, phishing, denial of service attacks, and listening devices [2]. Even unauthorised readers can endanger privacy by viewing tags without enough access restrictions. Due to anticipated tag responses, even if the tag content is secure, it may still be tracked; a traffic analysis attack may jeopardise "location privacy." Denial-of-service attacks by attackers pose a risk to the security of systems that rely on RFID technology. The remainder of the study classifies RFID tags according to their computing capabilities and discusses the issues of privacy and authentication individually.

3.1.4 Basic RFID tags

Since there are just a few thousand gates, most of which are used for rudimentary functions and extremely few of which are used for purposes of security, basic RFID tags cannot, as far as we are aware, execute actual cryptographic operations. In the belief that Moore's law will soon increase the power of inexpensive tags on the computational front, a lack of computing resources is considered a transient condition of affairs. Due to its widespread use, cost aspect is still a problem. Additionally, if barcodes are replaced, the price of such things will increase significantly. Therefore, we use alternative, less expensive methods to handle security and privacy problems here, such as the tag-killing tactic, specific blocker tag, re-encryption method, and numerous others.

3.1.5 Confidentiality

The different suggested solutions to the privacy issue are described in this section.

3.1.5.1 Tag approach to sleeping and killing

Killing [2] RFID tags on the things that are purchased is one of the simplest methods for protecting consumer privacy. The kill order may be interpreted as a time-consuming procedure that gradually renders the tag physically inoperable, perhaps by severing the antenna or blowing a fuse. The following kill attributes apply to tags supported by the Auto-ID Center [2]. Every tag contains a different 8-bit password, and once the password is received, the tag self-destructs. Although this method is helpful in preserving user confidentiality, using it requires conscious action.

3.1.5.2 Blocking

A plan for call blocking that uses "selective blocking" by "blocker tags" as a protecting personal data measure. Blocker tags are unique RFID tags designed to stop unauthorised screening of tags labeled within the privacy sector. JRS proposes that the tag will be blocked rather than being used for killing or sleeping.

3.1.5.3 Soft blocking

It [3] suggests an alternative to blocking that they refer to as soft blocking. Software (or firmware) modules that provide a different composition of functionality than standard blockers, as opposed to interfering with singulation, make up a soft blocker tag. In a nutshell, soft blockers only inform RFID scanners of their owners' privacy choices.

3.1.5.4 Relabelling approach

This method involves replacing the tag's original unique identification with a new one. But the previous one is still there and can be used. There have been many works based on this concept, such as the proposal made to overwrite RFID tags with fresh random numbers at every check. It offers a method for secretly screening library books. There are two strategies recommended for RFID tag. To conceal the tag's permanent ID under a user-provided private ID. In a different strategy, the tag's permanent identification is divided into two parts; a portion of the ID sequence is allocated to an object, and the remaining portion is provided by user-assignable RFID tags.

3.1.5.5 Re-encryption

This technique's main notion is to alter the appearance of ciphertexts while preserving the integrity of the plaintexts, or encrypted serial numbers, that are used as their basis. When employing RFID tags, re-encryption is a technique used to protect user privacy. They base their technique on a single key pair and an arbitrary public-key cryptosystem. An appropriate law enforcement organisation's private key SK and public key PK. An RFID tag in the JP system is carrying the specific identifying S. The ciphertext C produced by PK's encryption of S is released via the RFID tag.

3.1.6 Authentication

Authentication is a major issue with fundamental RFID. Because Class-1 Gen-2 EPC tags lack any specified anti-counterfeiting measures at all. A hacker might theoretically take the EPC from a target tag and put it into a phone tag, or they could imitate the target tag in another type wireless device. This section reviews some of the work regarding this issue. In [4], Juels puts out the idea of a yoking-proof (using the definition of "yoke" as "to join together"). It serves as evidence that two RFID tags were scanned concurrently. This method's specific goal is to enable tags to produce a proof that may be verified offline by a reliable party, even while readers are perhaps suspect.

3.1.7 Other techniques

In order to defend against numerous assaults like skimming, spoofing, and cloning, Sun and Ting [3] present an RFID authentication system based on a gen2 tag. This scheme's fundamental tenet is similar to other pseudonym-based ones in that it provides mutual authentication between legitimate readers and tags while randomly generating every session. To the authentication, this uses the PRANG and CRC-16 functions. But in this case, additional circuitry is required to determine the hamming distance between two 16-bit values. Therefore, hamming distance measures are used for this. In order to provide stronger security without connecting to a back-end server, present a lightweight protocol that relies on threshold secret sharing and key transportation for authentication. It guards against information tracking, listening in, and active information leakage. This scheme's key benefit is that it may be used with inexpensive RFID devices. Length is still appropriate for passive tags.

3.2 WIRELESS SENSOR NETWORKS IN MANUFACTURING SYSTEMS

During the production it is mandatory for maintaining the quality, efficiency, so that the research on RFID-based wireless manufacturing systems has lately been conducted in a variety of areas. Recently, a number of strategies for generative RFID systems have been proposed. This section categorises the various study subjects in order to evaluate their contributions to the field of research.

3.2.1 Impeccable timing

Manufacturing impeccable timing is introduced to enhance competition of manufacturers by cutting lead times and inventories. Several issues, including continuous and rapid-time data sharing and the production plan should be adjusted in response to the manufacturing environment's ongoing changes through communication between various functional departments be examined and resolved in order to implement impeccable timing in the manufacturing systems. The impeccable timing systems can be modified with the help of the RFID to create manufacturing techniques. The smart RFID technologies can be used in the food industry's impeccable timing process to supply raw ingredients, increase consumer engagement with manufactured goods, and improve display quality [5].

3.2.2 Lean manufacturing using RFID

Lean manufacturing systems are used to cut back on waste throughout the part manufacture process. By monitoring the movement of goods and supplies in production techniques, RFID can help to establish lean manufacturing systems and lower the cost of manufactured parts. To eliminate waste materials in the manufacturing system, the part production process investigated RFID systems inside a lean supply chain in a global setting [6].

3.2.3 Wireless manufacturing in real time

The indoor positioning system powered by OS-ELM utilises an RFID system to reduce time on training in part manufacturing technologies by obtaining actual production data. Figure 3.1 depicts the algorithm developed for the OS-ELM-based indoor positioning system [7], accessing helpful data on items both during and after the manufacturing operation.

3.2.4 Systems for handling materials

Material handling expenses in part manufacturing techniques, which make up a sizeable portion of inventory cost, can be decreased with the use of RFID devices. Figure 3.2 depicts real-time management with RFID [8].

In order to reduce the amount of time required for material handling between the suppliers of automotive parts and the parts' manufacturer, the material management system for automobile parts received RFID technologies for the first time. To cut down on the time and expense of material flow in part manufacturing systems, researchers are studying and optimising material delivery systems.

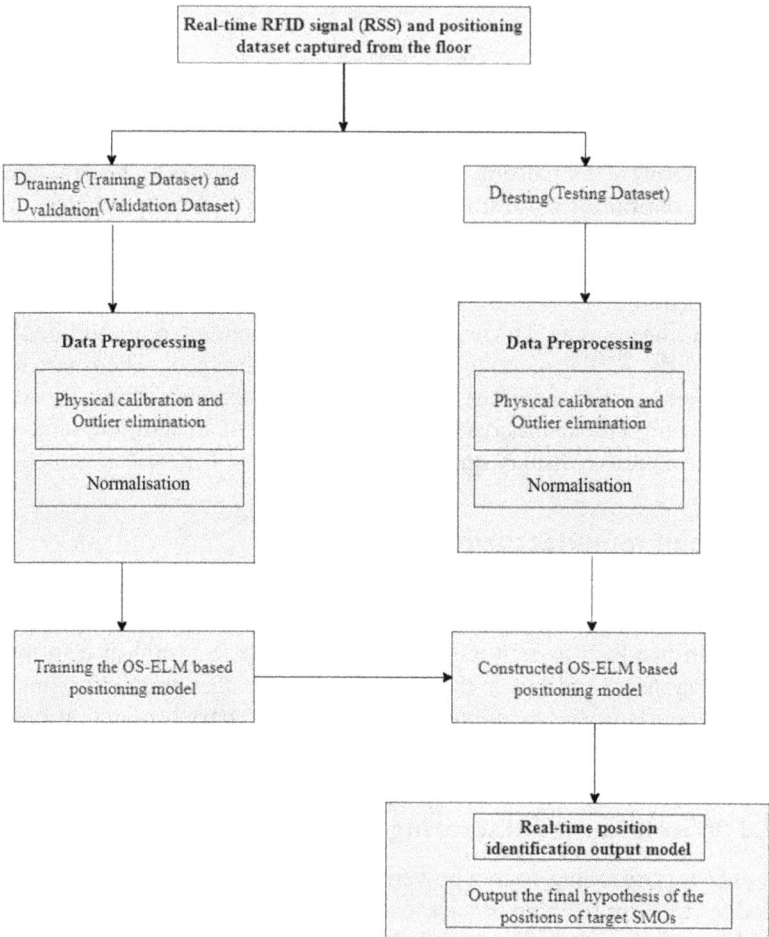

Figure 3.1   The indoor positioning system's created algorithm is based on OS-ELM.

Source: Yang et al. (2016) [7].

3.2.5 Planning the assembly of the product

Using RFID to enhance the effectiveness of the element production chain, the planning of electronic parts assembly for big air conditioner systems. In order to provide enhanced material and labour flows in the production process, RFID systems are used in the assembly operation of fixed-position layouts while accounting for the space available area. The study's created wireless manufacturing systems are shown in Figure 3.3 by Huang et al. [9].

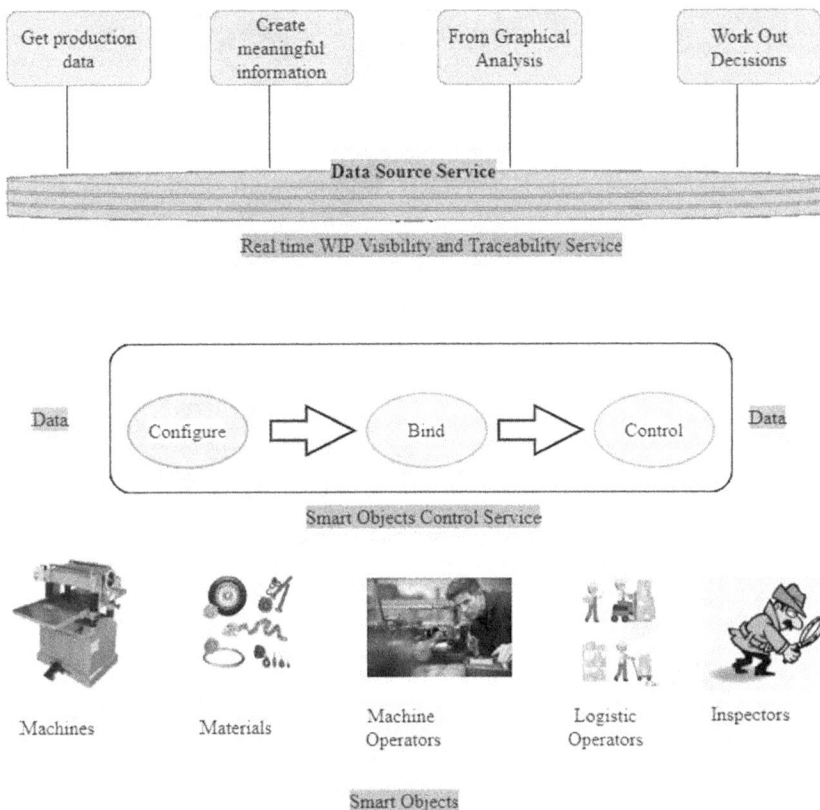

Figure 3.2   Real-time management with RFID. Source: Zhong et al. (2013) [8].

By employing mobile agents and RFID technology to monitor and control the complex product assembly process, it offers a solution to the problem of asynchrony between the logistics stream and the information stream in the executive process of complicated product assembly.

3.2.6 Management of the supply chain

RFID technology can be used to enhance the process of making decisions for part production and improve supply chain management. In order to increase the efficiency of production systems, RFID systems can be used to build a variety of distribution network modules, such as inventory management, investment management, storage operation, manufacturing processes, and merchandise sales. It provides a literature review on the expanding subject

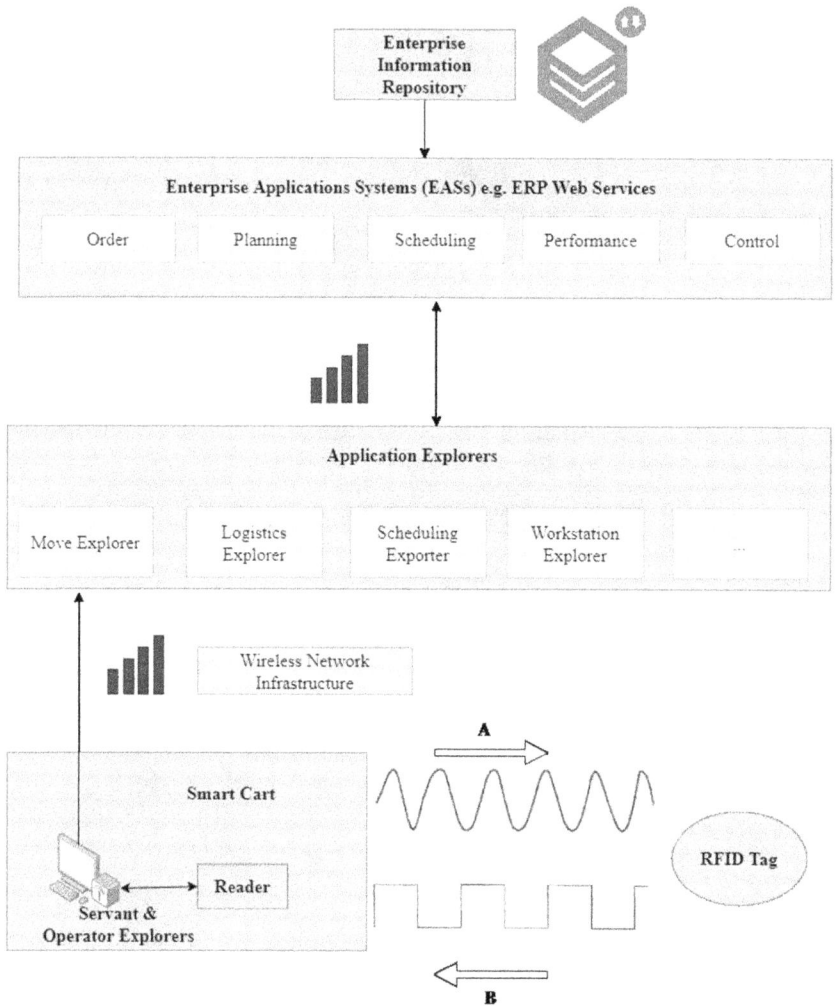

Figure 3.3 Conceptual architecture of wireless manufacturing. Source: Huang et al. (2007) [9].

of IOT employing RFID and its applications in supply chain management to establish an automated and networked process automation environment [10].

3.2.7 Utilisations of RFID in industry 4.0

By offering smart products and digital factories, industry 4.0 represents the next stage for the manufacturing sectors. To improve the capabilities and adaptability of smart factories and intelligent production systems, RFID technologies can be used in industry 4.0 [11].

3.2.8 Wireless sensor networks in manufacturing systems

Recently, various research works have taken into consideration the deployment of a huge population of sensors in the part production environments for comprehensive monitoring and regulation. The use of wireless sensor networks to track and assess manufacturing equipment health. A process for monitoring the manufacturing environment with wireless sensor networks along with the Internet of Things to enhance resource utilisation and energy consumption [12].

3.2.9 Energy management in networks of RFID-sensors

The development and design of energy management systems is focused on maximising energy utilisation in industrial systems. RFID-sensor systems can be used to build energy-efficiency systems to improve the efficiency of energy use in manufacturing operations.

3.2.10 RFID-based production scheduling and planning

Recently, methods for production planning and scheduling were created employing precise data from RFID systems during the part manufacturing. As a result, using RFID technology during the part production process can result in greater added values. The geofencing algorithms and RFID devices to integrate a system that makes decisions intelligently in the part production technique [13]. The use of RFID technology to improve the efficiency of casting-process part production in steel manufacturing enterprises.

3.2.11 Performance enhancement of manufacturing systems

The data gathered from the RFID systems is used to measure and alter the performances of part manufacturing systems.

For optimality and availability, an intelligent resource allocation in the IIoT paradigm. IIoT typically comprises of a variety of wireless sensors, including RFIDs, for remotely monitoring different techniques that are rapidly associated to real-world things used in company devices [14].

The concepts of ubiquitous computing in each nation are compared in Table 3.1 [15]. In order to convey the quantity produced amongst the participants by fusing the elements, a BM is to identify numerous components such as the products and services, business strategies and processes, and stakeholders of a BM.

Table 3.1 Ubiquitous comparison

Nation	US	Europe	Japan	Korea
Perception	Ubiquitous and prevalent computing	Vanishing and ambient computing	Ubiquitous network	Ubiquitous Application
Equity	Service by smart devices	Intelligent collaboration by information objects	Anyplace connection by tiny chip, intelligent card, framework roaming	Single function application using short assortment wireless Interface
Research domain	Computer devices	Every things	Networking	Application
Core technology	Sensor, MEMS, tiny size object chip			

At their core, all pervasive computing models (also known as ubiquitous computing) have a vision of small, low-cost, reliable networked processing units that are spread at all scales throughout daily life and are generally used for quite commonplace purposes. As an illustration, home-based purposes. Another typical scenario imagines refrigerators that are "conscious" of the appropriately-tagged contents and may design several menu options using the food they currently have on hand while also alerting users to expired or rotten food.

3.3 RFID ISSUES

3.3.1 Causes of unsuccessful RFID tag scans

Several factors, including tag detuning, tag collisions nearby the RFID system, can prevent readers from detecting RFID tags that are within their read range. The utilised playing cards with RFID tags to illustrate how some of these occurrences can lead to failed tag reads [16]. Because it accurately reflects the most common reasons for unsuccessful tag scans, they believed that the deck of cards situation is the ideal illustration.

3.3.1.1 Collisions of tags

Tags that don't transmit their IDs within the same time span are typically ignored. One exception to this rule is the capturing effect, which enables the reader to accurately recognise the data presented with one of the tags despite many tags reacting simultaneously. An opportunity exists in a stochastic generally pro-method where a tag might not be detected for at least one frame. Evidently, the number of available time slots and the availability of electronic tags have opposite effects on the chance of collisions.

3.3.1.2 Tag detuning

It explains how the antenna coil was connected to a capacitance in parallel by the tag manufacturers to create a constant frequency circuit [17]. The resonant voltage produced by the tuned tag also increases as readability during the resonance, considerably improving the outgoing range over frequencies. Because of the environment's detuning influences, which can also drastically reduce reading distance, its tag's resonant nature leaves it susceptible to them.

3.3.1.3 Additional sources of errors

The existence of metal inside the vicinity of the tag can also prevent successful reads because it bends the magnetic flow and decreases the power connection to the tag. If the tags are attached to a metal surface, they are typically tough to find. A reduction in read range is caused by metal near the reader antenna owing to antenna detuning, which is similar to tag detuning. As the label is turned in respect to the field lines, the coupling falls off until the tag is no longer recognisable.

3.3.2 Possible solutions

Even if a cryptoanalyst is aware of the entire cryptoanalytic process, breaking the cryptosystem requires incredibly vast quantities of computational resources and time to evaluate the data. The ability to adjust a small number of parameters and have a significant impact on the system as a whole is one of the prerequisites for this cryptosystem. Therefore, floating-point computations and other difficult numerical operations should be avoided while designing a cryptosystem for RFID.

3.4 SECURITY AND PRIVACY ISSUES IN RFID

Some security solutions, including as transaction concept, improved shadow reads, layered reading constraints, protected computation, and improved protected procedures, have been developed to support many tags reading safely and confidentially. There are, however, few ongoing privacy issues at the user and application levels. They must pay attention to a security flaw they are dealing with when creating and monitoring a "random number" key in order to prevent potential communication leaks or breaches of privacy caused by monitoring connectivity.

3.4.1 Security

The main barrier to implementing any new computer paradigms is security. The three basic security characteristics that have historically been used to classify risks to information systems are confidentiality, integrity, and availability. Despite the fact that they may not be mandatory to cover entire ideal protection characteristics of computerised system, these are a decent place to start when figuring out what is most crucial. The main issue is availability, particularly when a system combines several separate tasks. For instance, because the college's computer system oversees staff salaries, course scheduling, and library catalogues in addition to student grades, it cannot be offline for more than a week at a time.

3.4.2 Security issues in ubiquitous computing

The infrastructure for ubiquitous computing is a connectivity system that is portable. Because of the significant impact this design has on how consumers engage with ubiquitous services and apps. The infrastructure for mobile and wireless devices in the world of ubiquitous computing is fraught with many problems. For the purpose of determining the precise role and requirements, we believe that additional research is required in this area.

Figure 3.4 depicts the different security concerns in ubiquitous computing. The requirements could include assistance for dependable communications, additionally or combined access to numerous wireless communication networks. As individuals, organisations, or groups are hesitant to transmit sensitive and private data, and security is the fundamental issue with ubiquitous computing as it involves the transmission of mission-critical data through an environment that is neither secure or is not acknowledged as being secure.

3.4.3 Fake tag ID issues

The implementations used by Generation 2 RFID standards ensure that receiving tag ID is indeed a legitimate ID and not just noise or errors. This occurs when a signal processing software translates noise as a tag ID, was a significant barrier to the adoption of this useful technology. Ghost read is one of the issues with Generation 1 RFID technologies. In reality, the half-duplex mode of communication is still used. When the tag reads the orders from the reader, it doesn't talk; instead, the time limits reply in a set amount of time.

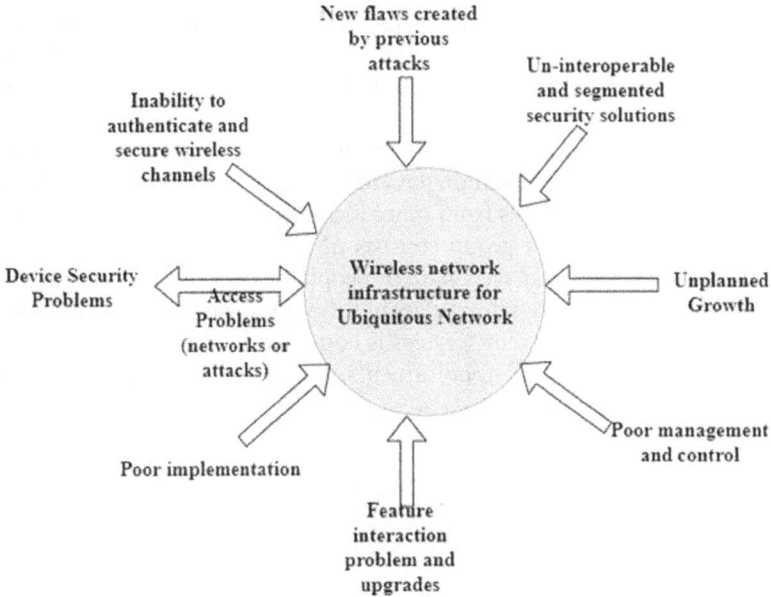

Figure 3.4 Security issues in ubiquitous computing.

3.4.4 Password protection and effective randomness

To prevent data from being sent through an air interface, the reader and tag must always maintain a secure connection. An 8-bit password is used in Generation 1: Class 1 RFID standards to carry out the "kill" instruction, which secures the data. Due to the 8-bit password's limited set of 256 possible values, it is neither secure nor difficult to crack. A 24-bit password is used in Generation 1 Class 0, which offers higher protection against unauthorised access to the information. Because of this, RFID has never before experienced a level of secure communication [18].

3.4.5 Location privacy

"Location-based services" is one of the trendy terms associated with ubiquitous computing. Due to the availability of user location data, ubiquitous computing systems offer geographically specific solutions. However, the location is a crucial amount of data, thus disclosing it to a complete stranger could compromise your security and privacy [15]. For instance, when walking home at night, people seek to conceal their position in order to reduce the chance of being robbed. Although the location data has collected a lot of attention, the requirements for access management have not been adequately assessed.

3.4.5.1 System for tracking people

The system for tracking people hierarchical structure is depicted in Figure 3.5. The arrows in the graph show which services communicate with which other services or devices, while the nodes in the graph are either devices or services.

Multiple location services are used to form the location system. Each location service either collects location data using a specific technology or analyses location data that it receives from other location services. It can be separated into two groups, the first group consists of services like device and person locator, and Cclendar that are aware of people's locations. The second group possesses services that are knowledgeable about the whereabouts of devices.

Information from different services is compiled by the people locator service. According to people's appointments, the calendar service detects their

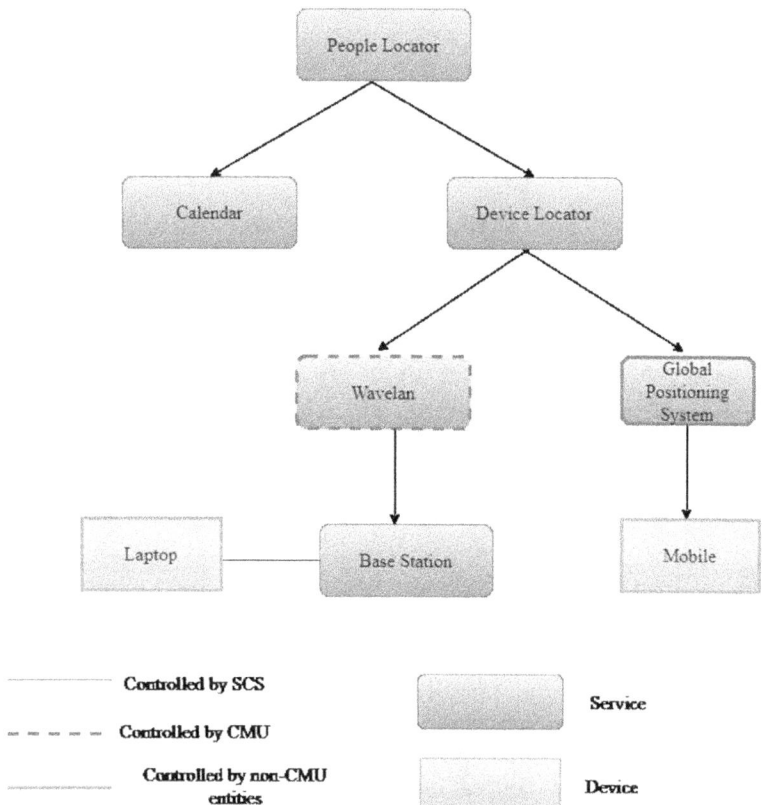

Figure 3.5   System for tracking people.

current location. The services of GPS and Wavelan make up this collection of services in Figure 3.5. The location is obtained by the GPS service from mobile devices with GPS capabilities. By identifying each wireless device's base station, the Wavelan monitors the track of its location.

3.4.5.2 Issues with location privacy

The most effective technique to safeguard position privacy is access management, which gives you the option of granting or denying access to your location data depending on the requester's identification and other details regarding the specific inquiry being made. Despite the fact that this security is effective, it is simple to identify anyone present at the place who is not the owner of the property. To resolve the issue, straightforward heuristics techniques [19] are also employed occasionally.

3.4.5.3 Location guidelines

Both individuals and services may ask for directions. A user's location or the persons in a particular space in a building, can be requested with a query (room query). Based on the two core searches, more intricate inquiries or services that provide location-specific information can be created. To prevent unauthorised parties from discovering your location, we could implement location restrictions. According to the following requirements, an access control programme for a system to track people must provide the following features:

- Granularity
- Locations/users
- Time intervals
- Institutional versus user policies
- Transitivity of access rights
- Conflicting policies

3.4.6 Example environments

In this part, we examine the definition and application of location regulations in two different contexts—a hospital and an academic environment.

3.4.6.1 Hospital

A multilateral security paradigm [15] that safeguards the information is communicated within the building secures medical data, for instance, as patient details. For instance, only certain hospital professionals who are concerned about a patient can manage their records. Locational data also requires a similar paradigm. The doctor merely needs to find the patient. A patient must also permit others to search for them, such as a spouse. The hospital will implement central regulation policies to fulfil these objectives. They may now have the authority to add more individuals to the user's policy. To safeguard patients' privacy, room policies should be created by the central authority. User and room systems do not need to be synchronised in the hospital scenario.

3.4.6.2 University

Students and teachers develop their consumer policies in a university context. However, a centralised authority set the rules for the room. The authority will set the room regulations for lecture halls and other spaces so that user and room policies must coincide. This means that the location system first verifies a user's identification by checking the room's or hall's user policies when the room got a request. User and room regulations are not coordinated for the workplace in this case. In this case, user regulations might be transitive but room regulations might not.

3.4.7 Authentication

Students and teachers develop their consumer regulations in a university context. However, a centralised authority set the rules for the room. The authority will set the room regulations for conference halls and other spaces so that user and room policies must coincide. This means that the location system first verifies a user's identification by checking the room's or hall's user policies when the room receives a query.

3.4.7.1 Confidentiality

The security risks for ubiquitous computing fear are no longer valid because we now have mature and reliable symmetric cyphers for preserving the confidentiality of a communications channel. The real issues lie elsewhere, new restrictions are imposed by the average ubiquitous computing device's size and design. Since untethered gadgets run on batteries, they are unable to utilise the newest and most potent processors. They need very regular recharges in order to prevent such issues (as laptop users know all too well). As a result, the processors in many ubiquitous computing devices are too slow for computationally demanding applications like public-key cryptography.

3.4.7.2 Integrity

The main integrity issue is preventing malicious third parties from tampering with messages sent between two parties. Similar to confidentiality, this issue may be easily solved utilizing well-known cryptographic techniques such message authenticating broadcast data is a bit difficult, but academics have developed a number of chaining protocols to address this issue. Therefore, the device itself rather than the communications that are being sent or received in it presents the most critical integrity issue for ubiquitous computing.

3.4.7.3 Availability

Jamming the communication channel is how wireless systems are typically attacked to reduce their availability. Although jamming can cause short-range RF communication systems used in ubiquitous systems to fail, there are methods for dealing with it that are outside the scope of system design. Additionally, the connection between security and energy saving raises the likelihood of a denial-of-service attack. Keeping a gadget awake until its battery runs out can be an efficient and targeted assault if it has a limited amount of battery life and is trying to sleep as much as possible to conserve it. When the victim's battery runs out, the assailant can simply leave after disabling them. This violent approach can be described as sleep deprivation abuse.

3.5 SECURITY PROBLEMS IN UBIQUITOUS COMPUTING: SECURITY FLAWS, THREATS, AND SOME DEFENSIVE SYSTEMS

3.5.1 Security flaws

3.5.1.1 Network dynamics

An attribute of ubiquitous computing environments is their lack of permanent infrastructure, centralised servers, centralised power, and centralised trusted third parties. A network must be able to self-configure due to the flexibility of portable devices to join or quit networks.

3.5.1.2 A large number of nodes

The second most common issue in an ubiquitous computing system is brought on by the large number of nodes involved in network connections. Due to the fact that certain nodes may behave maliciously (attempt to disrupt network operations), selfishly, or with a dynamic style. If we don't manage with care, the entire network could come crashing down. Therefore, managing bad behavior nodes in trust computation and management is a difficulty.

3.5.1.3 Resource constraints

Due to the devices' resource constraints, it is challenging to ensure service availability and implement security measures that depend on intricate computations. Authentication-related difficulties designing surroundings where individuals cannot be aware of the technology around them is the aim of ubiquitous computing. For instance, to protect the user's critical data from unauthorised access, mobile phone locking techniques like credentials or trend are necessary. Because of this, some people utilise the same password on many devices, while others store a list of passwords on their devices, giving their adversaries a chance to access confidential data.

3.5.2 Security threats

It discusses a few security breaches in the context of ubiquitous computing.

3.5.2.1 Cryptanalytic threat

This type of attack includes password cracking attacks, side channel attacks, electromagnetism attacks, auditory cryptanalysis key search attacks, and other cryptanalytic attacks for such general information purposes, cryptosystem, birthday, and key distribution attacks.

3.5.2.2 Denial of service (DOS)

This kind of assault seeks to deny the intended consumers of the services and resources. For instance, by destroying the network, degrading the services, and exhausting the resources of the device, or giving the nodes involved in communication erroneous trust ratings Therefore, it is crucial to take the proper action to discover these conditions beforehand.

3.5.2.3 Eavesdropping threat

This category of attacks includes a number of methods in which a malicious person listens in on conversations to learn about sensitive information or tampers through the communication medium by altering messages. The opponent discreetly peers over the shoulder of the victim to view the contents of the screen of a mobile device. Another illustration is spying, in which a foe can covertly watch what someone is typing on a computer.

3.5.2.4 Man-in-the-middle attack

Device authentication is crucial for providing services in a world where computing is ubiquitous. Mandatory authentication of the artefact using a secret, such as a password or PIN number, is required.

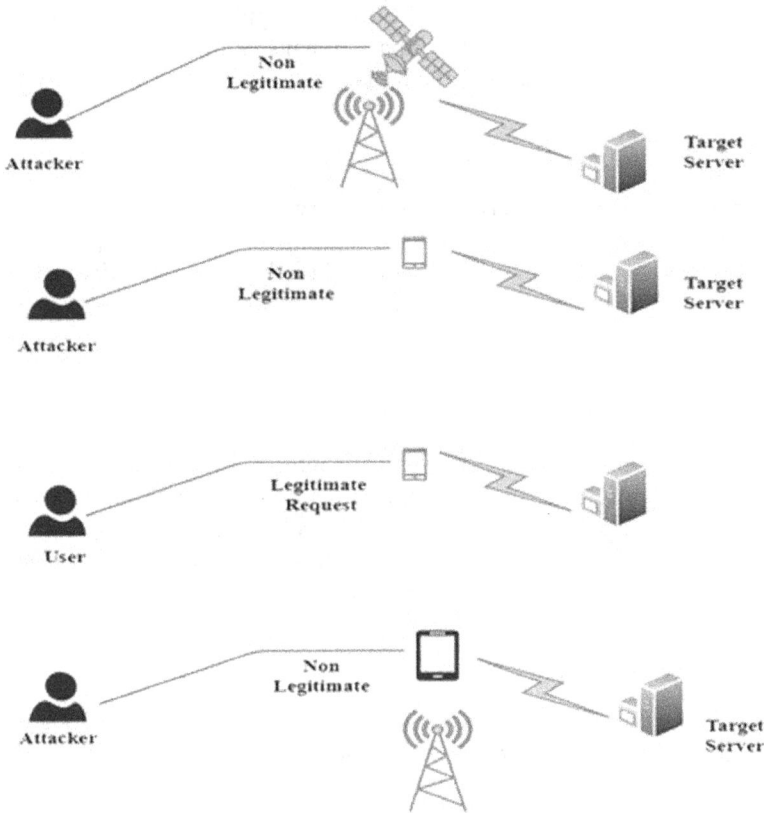

Figure 3.6   Attacks on different nodes.

When users or artefacts forward challenges and responses to mimic the presence of other actors, the man-in-the-middle attack takes place. Figure 3.6 depicts the attacks on different nodes in ubiquitous computing [15]. In other words, an attacker doesn't need the card and doesn't need to interfere with the terminal to change the transaction.

3.5.2.5 Risk to access network

The access network connects the networks of the external service providers with the home gateways. Evidently, critical data, such as accounting reporting, user IDs, and other specifics, may be publicly disclosed if an intruder at the home data connection point intercepts confidential data from network packets.

3.5.2.6 Unauthorised connection

The attackers also take control of the home appliances while posing as a legitimate internal customer. Information leaks can result in its misuse for purposes unrelated to those of the users.

3.5.2.7 Capturing sensitive data

Electronic sensors are widely used in the pervasive environment due to their limited computational resources in the monitoring process. By positioning a receiver next to a sensor, an attacker can directly collect sensitive data from it. Instead of cryptography matters, these sensors often focus on sensing duties.

3.5.2.8 Stealing intermediary device

Typically, the sensor data is collected by an intermediary device. The moment an attacker gains access to a device, that device cannot be used again and is considered to have been breached as a source of network information for attacking purposes. Since the equipment has a maintenance interface, this frequently creates a possible vulnerability.

3.5.2.9 Data manipulation

They use an intermediary device to maintain the traversing sensor log details since the sensors' limited processing capabilities prevent them from immediately authenticating the data that is being passed. Insiders and impersonation: in this form of attack, the attacker may substitute bogus devices for the real ones so that a hacker acting as you can obtain the network's sensitive data.

3.5.3 Some defensive systems

The main obstacle to the adoption of ubiquitous computing is security. In this section, we present possible solutions for these kinds of environments and networks in order to solve the potential security concerns.

3.5.3.1 Real-time intrusion detection

The current intrusion detection system (IDS) can also be put directly into ubiquitous networks due to a lack of consideration for their adaptability, heterogeneity, and resource constraints. The proposed solution to this problem is a provider and reader intrusion detection system, which logs events and recommends defence strategies for different infrastructure devices against intrusions [20]. SUIDS, in short, consists of the following:

- It offers the potential distributions for services.
- It calculates the statistical difference between past and present behaviour.
- It determines whether or not the conduct constitutes an intrusion.

3.5.3.2 RFID-based authentication protocol

A radio frequency identification (RFID) device is a microchip that uses radio frequency (RF) impulses to transmit a specific serial number and other data. By incorporating tags into the objects, RFID allows for remote object identification. Among other things, procurement, industrial, and inventory control all benefit from the use of RFID tags. In a setting where computers is pervasive, RFID readers or components are mobile. There are normally two possible values for an RFID tag's ID state: dynamic or static. Although RFID tags is perfectly adapted for connecting the real world and the digital one, there is still a considerable distance to go until it can be said to be used everywhere.

3.5.3.3 Role-based access control

The foundation of a role-based access control (RBAC) system is the many roles that individuals play inside an organisation. This approach grants each role a set of privileges that enable it to assume a position inside a hierarchy among many other things. Two of the mappings are user role allocation (URA) with role authorisation assignment, both of which are changed individually. It has been modified so that it can be used in environments that permit ubiquitous computing.

RBAC contains the following, in brief:

- It lists all of the rights connected to the user role.
- The user's typical behaviour is obtained.
- Role-based privileges are subject to allowance controls.

3.5.3.4 Information leakage

This issue is crucial to resolve because of the sensitive information, especially in pricey items, and user worry over their information security. RFID systems, on the other hand, only respond to queries pertaining to the neighboring domain with distinct emitting signals. Users may not be aware of information leakage when it is placed. When a property has a high value, information leakage by insiders is more problematic, and the issue is more serious if the leakage is detected through data sharing and information accessibility [4].

3.5.3.5 Cryptographic protocols

A new area of study called "lightweight cryptography" encryption that has reduced computational overhead as small and resource-constrained devices proliferate. Asymmetric, asymmetric, and asymmetric cryptographic algorithms make up the majority of protocols used in ubiquitous computing, with the majority of researchers concentrating on hybrid techniques to create light-weight protocols [21, 22].

3.5.3.6 Using trust in computing and management

Mobile nodes in a ubiquitous computing network can interact with other nodes on their own initiative. It gets harder to guarantee proper behaviour as there are more and more mobile devices functioning as network nodes. Nodes that are malicious are kept out of the network. Since nodes may grow to be uncooperative over time, it is crucial to maintain dynamics in the computation of trust by keeping an eye on their behaviour and recalculating the assigned trust values [23, 24].

3.5.3.7 Traceability

An enemy could intercept the communication delivered in accordance to something like a targeting tag and link the two together. An opponent can use this URL to find the user's location. When doing authentication, the following considerations must be made:

- Get the product RFID tag details.
- It is converted for use by the database server.
- Its authenticity is verified.
- It is compared to an entity that certifies tags.

3.5.3.8 Biometrics

Although less noticeable, it has the potential to offer reliable, automated methods for detecting and verifying identity. Techniques like face or finger print recognition are quicker than entering safe passwords and don't call for specialised equipment like PDAs. Unfortunately, unlike other means of verification, the biometric traits cannot be modified or replaced because they are a natural part of the individual [25]. As a result, they become compromised and useless after being lost or stolen. Although there is still no reliable means for securely storing biometric information, biometric authentication systems maintain anonymity while providing sufficient adjustability to allow for missing matches and reduce a reasonable level of confidence.

3.6 CONCLUSION

Among the most important problems with ubiquitous computing is security. It allows communication devices to move freely, wherever they are, whenever they choose. As a result, contemporary computer networks are now more commonplace. The main issue is security when services are readily available to all different networks and their consumers. This chapter covered security concerns in a setting where ubiquitous computing is prevalent, including location privacy, susceptibility, attacks, and some responses against attacks.

REFERENCES

1. Ahsan, K., Shah, H., & Kingston, P. (2010). RFID applications: An introductory and exploratory study. *IJCSI International Journal of Computer Science Issues*, 7(3), 1–7, preprint arXiv:1002.1179.
2. Pateriya, R. K., & Sharma, S. (2011, June). The evolution of RFID security and privacy: A research survey. In *2011 International Conference on Communication Systems and Network Technologies* (pp. 115–119). IEEE.
3. Sun, H. M., & Ting, W. C. (2008). A Gen2-based RFID authentication protocol for security and privacy. *IEEE Transactions on Mobile Computing*, 8(8), 1052–1062.
4. Garfinkel, S. L., Juels, A., & Pappu, R. (2005). RFID privacy: An overview of problems and proposed solutions. *IEEE Security & Privacy*, 3(3), 34–43.
5. Chen, L. F., & Chen, S. J. (2017). A RFID-based JIT application for least waiting time for dynamic smart diet customers. In *ITM Web of Conferences* (Vol. 11, pp. 03009). EDP Sciences.
6. Smith, A. D., Damron, T. S., Cockrell, S., & Melton, A. M. (2018). Radio frequency identification systems within a Lean supply chain in a global environment. In *Encyclopedia of Information Science and Technology, Fourth Edition* (pp. 5516–5526). IGI Global.
7. Yang, Z., Zhang, P., & Chen, L. (2016). RFID-enabled indoor positioning method for a real-time manufacturing execution system using OS-ELM. *Neurocomputing*, 174, 121–133.
8. Shankar, R., Sarojini, BK., Mehraj, H., Kumar, AS., Neware, R., Singh Bist A. (2021). "Impact of the learning rate and batch size on NOMA system using LSTM-based deep neural network", *The Journal of Defense Modeling and Simulation*. doi:10.1177/15485129211049782
9. Huang, G. Q., Zhang, Y. F., & Jiang, P. Y. (2007). RFID-based wireless manufacturing for walking–worker assembly islands with fixed-position layouts. *Robotics and Computer-Integrated Manufacturing*, 23(4), 469–477.
10. Naskar, S., Basu, P., & Sen, A. K. (2020). A literature review of the emerging field of IoT using RFID and its applications in supply chain management. In *Supply Chain and Logistics Management* (pp. 1–25). IGI Global Publisher. doi: 10.4018/978-1-7998-0945-6.ch096
11. Ascher, A., Lechner, J., Nosovic, S., Eschlwech, P., & Biebl, E. (2016, July). Localization of UHF RFID transponders regarding industry 4.0 scenarios using DoA estimation techniques. In *Smart SysTech 2016; European Conference on Smart Objects, Systems and Technologies* (pp. 1–6). VDE.
12. Kumaresan, M., Basha, M. J., Manikandan, P., Annamalai, S., Sekaran, R., & Kumar, A. S. (2023). "Stock price prediction model using LSTM: A comparative study," *2023 3rd Asian Conference on Innovation*

in Technology (ASIANCON) (pp. 1–5) Ravet IN, India, , doi: 10.1109/ ASIANCON58793.2023.10270708

13. Oliveira, R. R., Cardoso, I. M., Barbosa, J. L., da Costa, C. A., & Prado, M. P. (2015). An intelligent model for logistics management based on geofencing algorithms and RFID technology. *Expert Systems with Applications, 42*(15–16), 6082–6097.

14. Udayakumar, K., & Ramamoorthy, S. (2022). Intelligent Resource allocation in industrial IoT using reinforcement learning with hybrid meta-heuristic algorithm. *Cybernetics and Systems, 54*(8), 1241–1266. doi: 10.1080/ 01969722.2022.2080341

15. Kim, W. S., Kim, J. K., Kim, H. K., Kim, C. S., Koo, H. S., Lee, S. B., … & Kim, S. K. (2003). The technology, infrastructure and trend of Ubiquitous computing. *Korea Information Processing Society Review, 10*(4), 23–38.

16. Shoba Bindu, C., & Sasikala, C. (2019). Security in Ubiquitous computing environment: Vulnerabilities, attacks and defenses. In *Ubiquitous Computing and Computing Security of IoT* (pp. 101–127). Springer, Cham.

17. Kumaresan, M., Basha, M. J., Manikandan, P., Annamalai, S., Sekaran, R. & Kumar, A. S. (2023). "Stock price prediction model using LSTM: A comparative study," *2023 3rd Asian Conference on Innovation in Technology (ASIANCON)*, (pp. 1–5). Ravet IN, India. doi: 10.1109/ ASIANCON58793.2023.10270708

18. Kang, Y. B., & Pisan, Y. (2006, November). A survey of major challenges and future directions for next generation pervasive computing. In *International Symposium on Computer and Information Sciences* (pp. 755–764). Springer, Berlin, Heidelberg.

19. Beresford, A. R., & Stajano, F. (2003). Location privacy in pervasive computing. *IEEE Pervasive computing, 2*(1), 46–55.

20. Suresh Kumar, A., Jerald Nirmal Kumar, S., Chandra Gupta, Subhash, Shrivastava, Anurag, Kumar, Keshav, Jain, Rituraj. et al., (2022). IoT Communication for grid-tie matrix converter with power factor control using the adaptive fuzzy sliding (AFS) method, *Scientific Programming, 3*, 1–11.

21. Ramanan, M., Singh, L., Suresh Kumar, A., Suresh, A., Sampathkumar, A., Jain, V., Bacanin, N. (2022). Secure blockchain enabled Cyber–Physical health systems using ensemble convolution neural network classification. *Computer and Electrical Engineering, 101*, 108058.

22. Kirubakaran, J., Prasanna Venkatesan, G.K.D., Baskar, S., Kumaresan, M., Annamalai, S. (2020). Prediction of cirrhosis disease from radiologist liver medical image using hybrid coupled dictionary pairs on longitudinal domain approach *Multimedia Tools and Applicationsthis Link is Disabled, 79*(15–16), 9901–9919.

23. Ali, M. A., Balamurugan, B., Dhanaraj, R. K. & Sharma, V. (2022). "IoT and blockchain based smart agriculture monitoring and intelligence security system," *2022 3rd International Conference on Computation, Automation and Knowledge Management (ICCAKM)*, (pp. 1–7). Dubai, United Arab Emirates. doi: 10.1109/ICCAKM54721.2022.9990243

24. Sureshkumar, A., Samson Ravindran, R. (2016). Swarm and fuzzy based cooperative caching framework to optimize energy consumption over multimedia wireless sensor networks. *Wireless Personal Communications, 90,* 961–984. https://doi.org/10.1007/s11277-016-3274-0

25. Sharifi, A., Khosravi, M., & Shah, A. (2013). Security attacks and solutions on ubiquitous computing networks. *International Journal of Engineering and Innovative Technology (IJEIT), 3*(4), 1–6.

Chapter 4

Privacy and security assurance in order to adopt a substantiation protocol in ultra-lightweight RFID

G. Sathya¹, M. Sathiya¹, P. Sumitra, S. Sabitha¹,
S. Bhuvaneswari¹, and Ramesh Sekaran²

¹PG and Research Department of Computer Science & Applications Vivekanandha College of Arts and Sciences for Women (Autonomous), Namakkal, Tamilnadu, India
²School of Computer Science and Engineering, Jain Deemed-to-be University, Bengaluru, Karnataka, India

4.1 INTRODUCTION

Wireless technology is now pervasive in our daily lives. The trend toward ever-cheaper, smaller, smarter mobile devices with lower energy usage appears to be continuing due to the ongoing advancements in microelectronics. These devices are often dependent on some finite energy sources since they lack the necessary cabling for the desired flexibility, which supports the need for power-aware design in both hardware and software development [1].

- A two-dimensional field function spatially sampled by a wireless sensor network (WSN). Only the communication links can be used by sensors (blue dots) for communication (dotted lines) shown in Figure 4.1.
- In this chapter, which is part of the National Research Network's programmer on signal and information processing in science and engineering?

  - Intelligent.
  - Wireless.

4.1.1 Integration of application

Three alternative interfaces can be used to incorporate project-specific application into the network system environment:

- The application might be created using a system template, which would then be saved in the system.
- All of the necessary functions to define and integrate an application component are provided by the KOM–OS operating system.
- An application can utilise the service-oriented messaging exchange system of the s-net services manager for data communication with other sensor nodes, services, or the spine component.

DOI: 10.1201/9781003362685-4

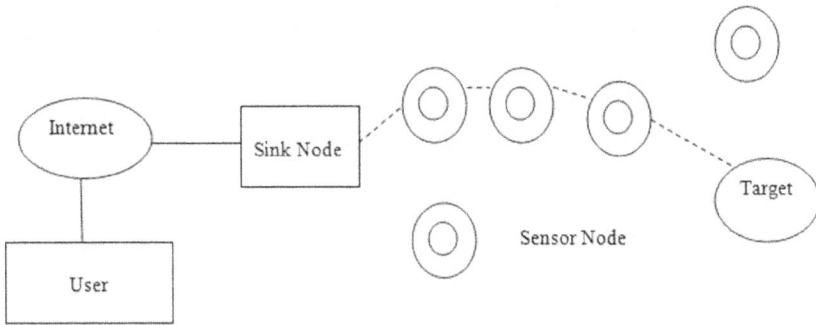

Figure 4.1   Wireless sensor network node connection.

4.1.2 Sensor nodes

- The role for establishing a connection between the wireless channel as well as a back-end infrastructure, such as an IP network, relies on opportunity nodes.
- Intermediate nodes are in charge of sending the data to the target for information delivery on a stationary basis (thus, "nodes"), such as for localization.
- For a wireless sensor there must be three logical categories of communication in an s-net sensor network. The key elements of the back-end system interface with the nodes' application components at the application server.
- Fundamental services or intricate procedures between nodes supplying the application components can speak with their function at the so-called middle ware level. The middle ware's components can speak with one another directly if necessary. The communication protocol regulates how messages are sent over the sensor networks.

4.1.3 Progression of RFID

Over several decades, RFID technology has gone through a number of stages as shown in Figure 4.2. The technology has been utilized in luggage handling, courier services, and tracking product deliveries. Other uses include automatic payment for all transactions, accessing control in departments of large buildings, control over both individuals and vehicles in a selected area, protection for materials that shouldn't be departing the location, development company equipment tracking, and hospital filing system [2].

Functional Complexity

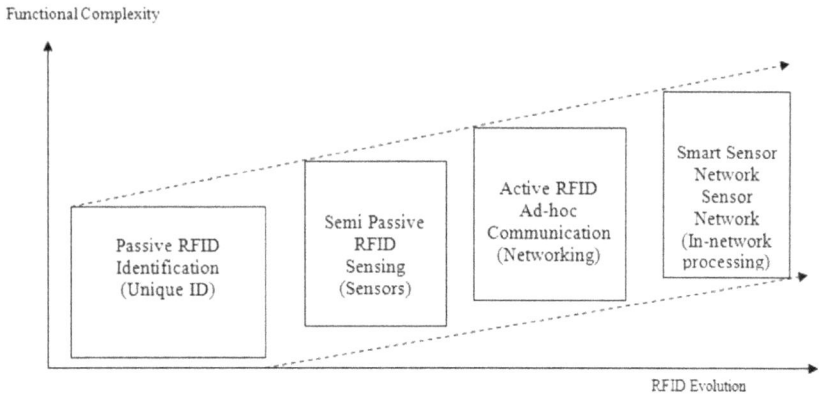

Figure 4.2   Progression of RFID.

4.1.4 In a functioning RFID system

The majority of RFID systems use tags that are affixed to the item being recognised. Depending on the type of application, each has a separate internal memory that is either "read only" or "rewrite". It is often set up to retain information on manufactured goods, along with the item's unique identification and manufacturing characteristics. The radio frequency identification system can find objects that are nearby thanks to the magnetic fields that the RFID reader produces. The tags reply to the query due to the reader's strong electromagnetic energy and the signal. The number of queries per second could reach 50. A large amount of data will be generated as a result. Industries involved in the supply chain manage this issue by applying filters to the backend information system [3].

There are various protocols to control the exchange of information among the reader and the tag. The verification process begins as soon as the reader is turned on. Specific frequency bands are where the protocol operates. When a tag enters the reader fields when reader is turned on, the device activates the signal automatically, and then responds by changing the field. The reader needs to recognise pointer impact in this scenario since all tags within the reader's range may respond simultaneously (indication of multiple tags). Conflict of a signal is avoided either by anti-collision algorithm's application, It, combined with the protocol utilised, permits the reader to sort and select the tag as a send on the frequency range (between 50 tags and 200 tags) [4].

4.1.4.1 An RFID system components

Various factors are combined to create an RFID system in the manner indicated in the section above. As a result, the RFID system may retrieve

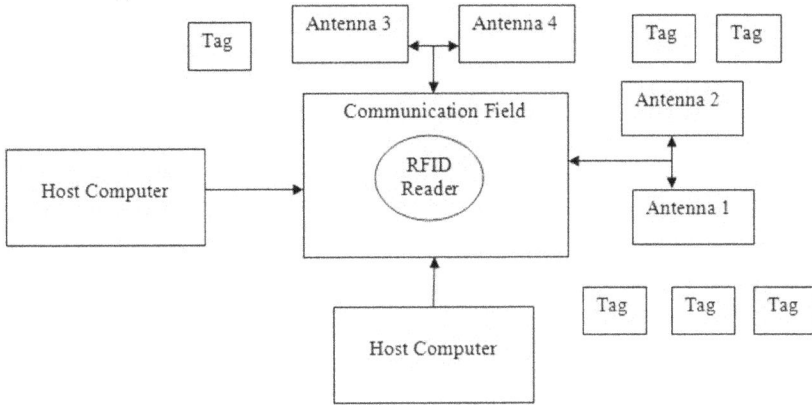

Figure 4.3   Objects in an RFID system.

the items (together with their tags) and perform several actions on them as shown in Figure 4.3.

- Tag (adhered to an item, individual identification).
- Antenna (tag detector, creates magnetic field).
- Starting to read more (receiver of tag data, manipulator).
- The RFID and reader capabilities' potential to interact with IT systems.
- Software with applications (interface, user database).

4.1.4.2 The secure RFID system concept with cluster verification

An RFID system (also known as a verifier) is built-up of wireless event labels (radio frequency reader, and a backend server). A tag is basically a silicon chip through projection and a little recollection that holds the EPC, which is its unique identifier (electronic product code). An object is often identified with a tag. The two main types of tags are active and passive. The power source for such active tags is located inside of devices. Large memory and complex processing capabilities are provided. These tags use an on-chip antenna coil that is driven by the reader's RF signal to communicate with the readers. As a result, they have relatively limited computational and communication capacities [5] as shown in Figure 4.4.

In various situations, dual or additional tags may be used toward collectively classify an objective. A technique that prevents grouping must demonstrate the simultaneous presence of two or additional radio frequency identification tags inside the transmission collection of the reader. The grouping-proof methods can be used either online or offline depending on

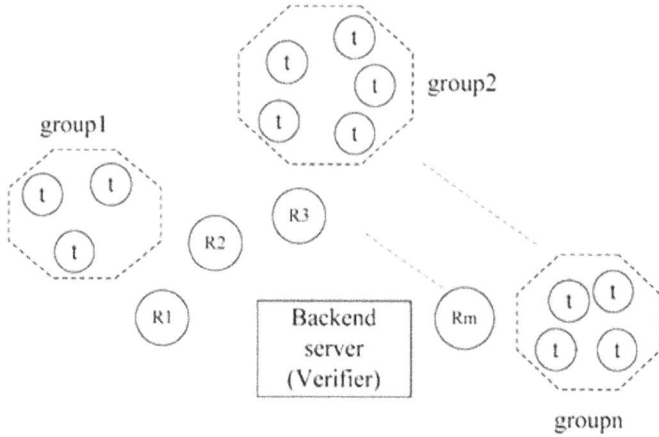

Figure 4.4 The radio frequency identification system components.

the verifier's function throughout the grouping-proof period. For the first mode, over the course of the protocol execution, the verifier container transmit and accept communications from particular tags (through the reader). On the additional pointer, once operating in offline mode, the verifier is just able to transmit challenges to the reader and is not required to remain present throughout the grouping-proof historical. Several grouping-proof protocols in use currently use disconnected mode. Two different categories of grouping proof procedures are presented: (1) serial mode and (2) parallel mode, depending on the condition for tags to finish signing within the grouping-proof period [6].

4.1.5 Protocol for ultra-lightweight mutual authentication

Malicious users are prevented beginning toward the inside of the system over the perception layer by the node verification tool used during the proof of identity procedure. Four categories for authentication protocols are as follows:

- These protocols are heavyweight because they use traditional cryptographic tools similar to hash functions as well as both private and public encryption.
- Middleweight: first protocols that maintain one-way hashing algorithms with fake random numbers manufacture reduction within this category.
- Inconsequential: these procedures can maintenance CRCs and other insubstantial pseudorandom number generators, among other lightweight features.

- Extremely lightweight: only simple bitwise logical functions are allowed to be incorporated for protocol creation in this class.

A connection between the EPC classes and the protocol classification is supported by some well-known examples. In order to save expenses, limit the silicon-based region of the EPC tags. An EPC C1G2 identifier usually has a response barrier of 32 K bits and can handle up to 4 K gate equivalent (GE) for cryptographic operations. The area needed to fabricate a two input NAND gate corresponds to one gate. Therefore, less GE for the implementation of the authentication protocol translates to lower overhead costs for security-based operations [7].

4.1.6 Protocol security analysis

Currently, the following are the main types of attacks used to compromise the RFID authentication protocol: attackers attempt to compromise the secrecy, integrity, and availability of the mutual message, for starters; another issue is that an attacker might use protocol flaws like resynchronization assaults, replay attacks, disclosure attacks, and more to mount attacks. We go over the security of our protocol as follows: mutual identification and data privacy: due to the fact that only legitimate readers and tags exchange secret values [8].

4.1.7 Protocol performance evaluation

According to the security premise, measures for determining if a protocol's performance is satisfactory include computational cost, communication cost, and storage requirement. The performance of our protocol is contrasted in Table 4.1 with a few popular ultra-lightweight authentication techniques that have recently been presented. The approach suggested in this chapter provides strong security-based performance. Calculated cost: the protocol solely uses straightforward bit-wise operations, including XOR and modulo addition [9].

Additionally, we use non-triangular operations like ROT, mix bits, and per. The only right shift and modulo addition operations needed for the mix bits operation and the only permutation operations needed for the per operation. As a result, even with their modest price tag, these operations are simple to carry out. In contrast to SASI and RAPP protocols, our protocol uses a pseudo-random number generator only once during a session of interaction, saving both time and processing resources [10], as shown in Table 4.2.

4.2 REVIEW OF THE LITERATURE

The most recent technical research on the concerns of privacy and security for RFID is discussed by the writers Ari Jules (www.arijuels.com). FID

Table 4.1 Comparison of barcode and RFID system

S. no	Barcode	RFID
I.	Cannot be read from a distance since they rely on the user to communicate with the reader	Can be read from a distance because they don't require the reader to interact with them
II.	One card may only be read at a time using barcode	It is allowed to read more than one at once
III.	Since embedded data cannot be uploaded, it is not permitted to repeatedly write the election data embedded in each card	Data embedded can be modified. This enables the repeated overwriting of each card's embedded electronic information
IV.	It prevents the activation of modern technologies, such as surveillance cameras, when an employee is nearby	RFID has made it possible for technologies, such as security cameras, to turn on when an employee is nearby
V.	It operates more slowly and needs time to focus	RFID requires no line of sight and is faster
VI.	It has a transponder with less data storage	It has a transponder with high data storage
VII.	This transponder is fake and inapplicable to tiny objects	The transponder is small enough to be integrated into other products

tags are small wireless devices, which are used to identify people and things. Because of the decreasing costs, they will probably continue growing into the billions in the next years. Due of its ability to track objects, RFID tags are being incorporated into consumer pockets, belongings, and sometimes even bodies. In supply chains, as well as discussing the social and technological backdrop of their work, scientist's proposals for privacy protection and integrity assurance in RFID systems are also included here. Any radio frequency device whose primary purpose is the identification of an object or a person is referred to as "RFID" the former definition [11].

As a new discovery to assist anonymous RFID authentication satisfies the requirements concerning privacy and availability, generic compilers are implemented to modify any challenge respond RFID authentication protocol into another one that performs key look up operations at a constant cost. They additionally demonstrate that, at least when the volume of tags is asymptotically enormous, any RFID authentication method that simultaneously gives assurances of privacy protection and of worst-case constant-cost key search must also suggest, public-key obfuscation. Also take into account relaxations of the privacy requirement and illustrate that a

Table 4.2 Comparison of ultra-lightweight authentication protocols

	LMAP	MAP	EMAP	SASI
Resistance to de-synchronization attacks	No	No	No	Yes
Resistance to disclosure attacks	No	No	No	Yes
Privacy & anonymity	No	No	No	Yes
Mutual authentication & forward secrecy	No	No	No	Yes
Total messages for mutual authentications	4L*	5L	5L	4L
Memory size on tag	6L	6L	6L	7L
Memory size for each on tag on server	6L	6L	6L	4L

simpler methodology should be explored to reach constant key-lookup cost if limited link ability is to be authorised [12].

As future technology to assist anonymous RFID authentication satisfies the requirements for privacy and availability, generic compilers are used to convert any challenge respond RFID authentication protocol into a different one that performs key look up operations at a constant cost [13, 14]. They further illustrate that, at least when the number of tags is asymptotically huge, any RFID authentication scheme that simultaneously provides a guarantee of confidentiality and of one of the worst constant cost key look up must also entail, public-key obfuscation. Also take into account regulatory changes of the privacy constraints and demonstrate that a simpler approach scan should always be taken to accomplish constant key-lookup cost if a limited connect ability is to be accepted.

Reference [15]: An analysis of conventional methods, with an emphasis on passive RFID tags is used to explain the procedure, helping us to identify the need for RFID tag authentication. This article aims to summarize the many, both new and historical, strategies of RFID tag identification and to draw attention to their weaknesses. The following phase in this project is to provide a collection of suggestions for criteria that will deal with the issues with authenticating RFID tags for use in product authentication. The principal contribution from the study specifies the circumstances in which a passive RFID tag should be implemented when considering authentication. Future studies are expected to identify how to employ passive RFIS tags while considering whose nature is predictable and controlled, without the requirements [16].

Reference [17]: Gen2+, an innovative system for detecting authenticity based on Gen2, has been described for economical RFID tags. This protocol ensures backward compatibility by triggering on every Gen2 message flow. Reader-to-tag authentication is implemented via the multiple round Gen2+ protocol, which already utilises shared pseudonyms and the cyclic redundancy check (CRC). Gen2+, on the other hand, executes tag-to−reader authentication using the recollection read command established in Gen2. Gen2+ is hence more robust to cloning and detecting intrusions.

The EPC global Class-1 Generation-2 technology (also known as Gen2 or ISO18000-6C) has been approved for usage globally; however the identification of the tag (TID) is communicated in plain text, making various tag options accessible and cleanable, including asymmetric or asymmetric cryptanalysis, which have been proposed based on typical cryptographic protocols.

Some indirect materials appropriate for lightweight authentication mechanisms have recently been published. The message flow of these protocols, however, differs from Gen2's. Readers presently in use may not read new tags. Our main priorities are to enable legitimate readers and tags to effectively authenticate each other and generate a new individuality each session, similar to existing pseudonym-based systems [18].

Gen2+, a new protocol solely utilising PRNG and CRC-16 functions for authentication, has been developed in an effort to make the original Gen2 standard operationally appropriate. The indicated tag is adequate for affordable RFIDs because it doesn't employ a cryptographic function. The Gen2 and Gen2+ tags may both be read by the present reader. Without compromising the Gen2 protocol flow, Gen2+ offers a high enough level of security for use in real-life situations. Here, it has also been examined how many rounds are sufficient for practical deployment in addition to significant update phase.

Reference [19]: The authors discussed mostly the protection of such a SASI protocol, an ultra-lightweight RFID protocol that was newly created and claims to already have higher security than older protocols. The fundamental problem in this situation is also in which SASI does not attain reluctance to follow. All contemporary RFID protocols are created to resist tracking with the location encryption for biological RFID users that must not be compromised even though RFID tags are numerous and are usually constant evident to the human user.

In view of the fact that most tags are passive devices that receive their power from an RFID reader's signals, additional technical challenge for RFIDs is reducing the amount of computing necessary for tagging. The most primitive binary operation and numerical operations, such as XOR, OR, addition, rotation, and delta, were also utilized to build a collection of ultra-lightweight RFID security protocols.

4.3 SYSTEMATIC BACKGROUND CONCEPTS

The latest developments in sensor technology make it possible to establish a large number of sensor nodes to create wireless sensor networks as shown in Figure 4.5. These enhancements extend to size, fuel efficiency, wireless communication, and manufacturing costs (WSN). These networks are established by dispersing a lot of often cheap, tiny sensor nodes throughout an area of interest in order to collect data about one or more variables.

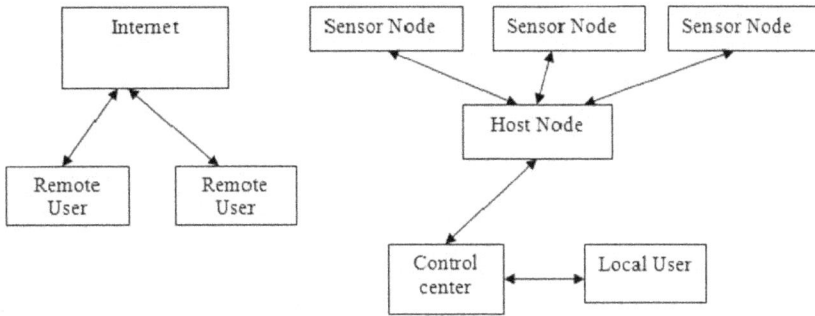

Figure 4.5   Overview of a wireless sensor network.

These nodes are primarily provided with the ability to perceive, process, and send data to other nodes and, ultimately, to the customer(s).

4.4 RADIO FREQUENCY IDENTIFICATION (RFID)

Some of the most universal internet equipment in history should begin to form, according to observers, in the region of automatic classification known as RFID. The mechanism of RFID is analogous to barcoding in its simplest basic form. It is recognised as being comparable to present technology and a method for improving computation. An electromagnetic distance authentication and communication transaction system is described in greater depth. RFID provides an enhancement over barcodes in terms of information density, two-way communication, and non-optical distance information exchange. RFID tags generally applied to readers gather the information from either the tags or objects or assets. Tags and readers are used by systems to link with objects (assets) and database systems in order to supply information and/or operational capabilities.

RFID *application characteristics.*

Four usage categories can be applicable to RFID devices:

- Positioning systems.
- Portable capture.
- Networked systems.
- EAS (electronic article surveillance).

4.4.1 Ultra-lightweight protocols for mutual authentication

The node authentication method used in the authentication process prevents unwanted users from breaking into the connection through the perception layer. The following are the four groups into which the authentication procedures were divided:

1. Heavyweight: these contributed to the persistence use of well-established cryptographic methods including private and public cryptographic algorithms and hash functions.
2. Middleweight: the protocols in this group only permit one-way hash functions and pseudo-random number generators.
3. Lightweight: these protocols are capable of supporting various light-weight techniques involving CRCs and other pseudo-random number generators.
4. Ultra-lightweight: for protocol construction, this class only permits the integration of straightforward bitwise logical functions.

4.4.2 Types of RFID tags

RFID tags may be categorized into active and passive variants. The first is active whilst the second is passive. A passive tag is sustained by a tag reader, whereas an active tag is sustained by an internal battery. By itself, a passive RFID tag won't have a power source or charger. It receives the support it needs from the reader. Thus, a passive RFID tag reader must be able to send out stronger electromagnetic signals and acquire comparatively weak messages from the passive RFID tag. The key distinctions between passive and active RFID tags are listed below in Table 4.3.

Table 4.3 Distinction between passive and active RFID tags

	Active RFID	*Passive RFID*
Source control	Residential (battery)	Peripheral (reader supplied) (reader provided)
Readability of tags	Capable of facilitating communications at a distance of up to 100 metres.	Only encompassed by the distance the reader occupies, which is frequently up to 3 metres.
Stimulated	A flexible tag is continually in use.	Only when a reader is present has an automated tag been activated.
Enticing field strength	Low since this tag uses a domestic sequence source for its signal transmission.	Elevated since this tag receives power from the reader's electromagnetic field.
Projection life	Limited to a sequence's life, which is around 5 years.	Extremely high, perfection doesn't end throughout a lifetime.
Record keeping	Storage space limitations; typically 128 bytes.	Storage space limitations; typically 128 bytes.
Cost	Exclusive	Reasonably priced
Size	Considerably compact (due to battery)	Smaller

4.4.3 System design

The combination of elements that are distinguishable from the alphabet's zero-symbol constitutes a string's Hamming weight. As a conclusion, it is comparable to the distance measure from the string constructed entirely of zeros. This is the quantity of 1s in the most real scenario, which is a string of bits. It is also known in this binary challenge, as the pop count, pop count, or the sideways sum. It is both a bit vector's l1 norm and the binary approximation of a given integer's digit sum.

Wei's work on extended Hamming weight for linear block codes and the work on secure network coding combine to expand expanded Hamming weight for linear block codes to linear network codes. In order to be more comprehensive, we discuss the network generalization Hamming weight (NGHW) for a linear block code in relationship to a specific linear network code.

We will introduce a network extension of the universal singleton bound in and demonstrate its tightness based mostly on NGHW for and through codes. Further, by constructing a linear network code that meets this constraint, we are able to recreate the process of creating a secure network code. Our network's generalization Hamming weight may accurately define the quality of linear network circuits on a telecommunication network, including secret sharing in classical cryptography as a specific case, by enhancing the original definition of the specialized Hamming weight.

4.4.4 Framed slotted ALOHA (FSA)

All of the competitive frames in the FSA algorithm have the same length. When a tag transmission and reception to a reader, the reader randomly generates a slot in the competing frame to receive the data, and if the data is successfully accepted, the reader will transmit an acknowledgement to the tag. Always be aware that in FSA, the reader utilises a limited contending frame size and the tag will deliver its data in a cyclic sequence up to the point at which the reader successfully gets the data. The FSA collision resolution algorithm's procedure. Let's say there are three tags that need to be determined, and there are the following four frame designs. The reader will first send a command to the tags instructing readers that the competing size of the image is four.

4.4.5 Simulation results

To evaluate the proposed scheme's performance in this domain, I implement the experiments. The experiments are designed by referring DFSA with DFSA and DFSA with optimum distance, and four distinct collision resolution techniques as shown in Figure 4.5. What follows are the essential

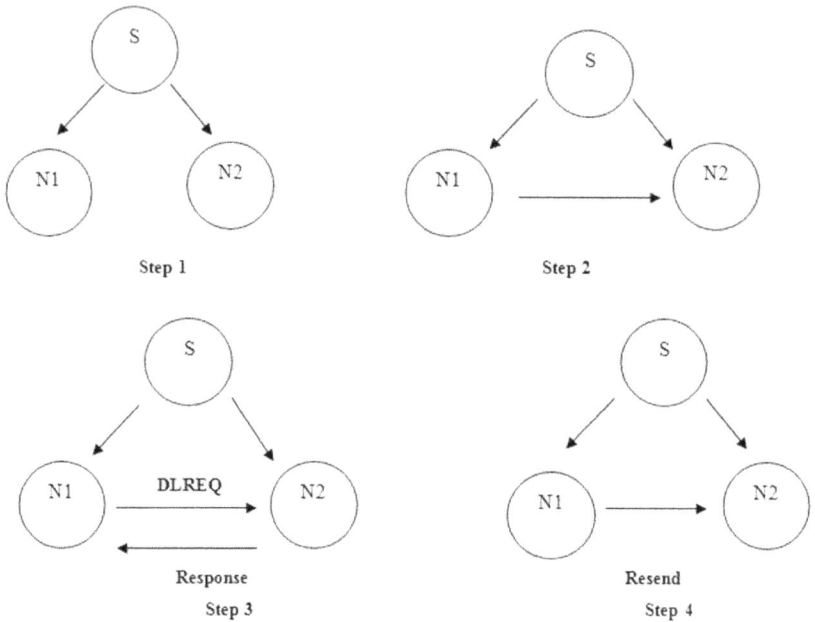

Figure 4.6   Intent communication system.

assumptions we adopted in our simulation in order to concentrate on system throughput-related constraints and to optimize simulations. First, all across the simulation no tags enter the reader's interrogation zone. Its shown in the Figure 4.6. Second, every prediction runs at least 1,000 times, and the default passwords used in the computations are indicated here. Finally, ideal frame size in DFSA specifies that the number of existing tags is equal to the optimal frame size. Addressing the proposed methodology, the frame size was established using the research outcomes. Additionally, DFSA and the recommended methodology will both change the competing frame size in the following ready cycle if the tag identification process is not yet accomplished.

4.4.6 Algorithm

Step 1: Server to connect node 1 and node 2.

Step 2: User to send the data original source to destination.

Step 3: To calculate the Hamming weight any data loss means that the current source node to send a request, provides one.

Step 4: Calculate data loss Hamming weight and then resend the data loss to the previous node.

$Hm = Pnid! = Cnid.$

Hm:Hamming weight. Pnid:Previousnodeid. Cnid:Currentnodeid.
Step 5: Calculate data without loss Hamming weight. Hm=Dnid== Csnid Hm:Hamming weight. Dnid:Destination node id.

Csnid: Current source node id.

4.5 CONCLUSIONS

The authentication mechanisms we established in this research for afford-able RFID tags are extraordinarily lightweight, and we also offered research on the protocols. Typically, the ultimate purpose of a method creator is to establish a protocol with provable security for a certain attacker model. Security reasons may draw from information theory (as in the case of a one-time pad) or they may restrict the challenge of cracking a protocol to a mathematical one. RFID has several advantages, many of which you may already appreciate to a large extent with the devices that are on the market. There are other benefits that are also relatively attainable, and it is believed that as technology advances, these advantages will eventually reach a mature state. But there's no denying that RFID offers a unique set of advantages that make it a potential instructor for a wide range of applications. Certain benefits may violate privacy rights, which could lead to complications when using RFID in specific circumstances. However, despite these problems, RFID is probably going to be the norm in other fields. RFID technology is changing rapidly now, and those modifications should be maintained.

REFERENCES

1. R.C.-W. Phan, "Crypt analysis of a new ultralight weight RFID authentica-tion protocol—SASI," *IEEE Trans. Dependable and Secure Computing*, Vol. 6, no. 4, pp. 316–320, Oct-Dec. 2019.
2. H.-M. Sun, W.-C. Ting, K.-H. Wang, "On the security of Chien's ultralight weight RFID authentication protocol," *IEEE Trans. Dependable and Secure Computing*, Vol. 8, no. 2, pp. 315–317, Mar-Apr. 2020.
3. P. D'Arco, A. De Santis, "On ultralight weight RFID authentication protocols," *IEEE Trans.Dependable and Secure Computing*, Vol. 8, no. 4,pp. 548–563, July-Aug. 2021.
4. M. Mallick, M.K.A.S. Govindaraju, A.S. Kumar, M. Kandasamy, P. Anitha, "Analysis of Panoramic Images using Deep Learning For Dental Disease Identification," *2023 Third International Conference on Artificial Intelligence and Smart Energy (ICAIS)*, Coimbatore, India, pp. 1513–1517, 2023. doi: 10.1109/ICAIS56108.2023.10073939
5. J.C. Hernandez-Castro, P. Peris-Lopez, R.C.-W. Pan, J.M.E. Tapiador, "Cryptanalysis of the David Prasad RFID ultra light weight authentica-tion protocol," in *Proc.2020 International Workshop on Radio Frequency Identification: Security and Privacy Issues*, pp. 22–34.

6. A. Suresh Kumar, S. Jerald Nirmal Kumar, Subhash Chandra Gupta, Anurag Shrivastava, Keshav Kumar, Rituraj Jain, et al., "IoT Communication for Grid-Tie Matrix Converter with Power Factor Control Using the Adaptive Fuzzy Sliding (AFS) Method", *Scientific Programming*, Vol. 03, 2022.

7. B.H. Jeong, C.Y. Cheng, V. Prabhu, J. YuB, "An RFID application model for surgery patient identification", *IEEE Symposium on Advanced Management of Information for Globalized Enterprises, AMIGE2008*, September 28–September 29, pp. 304–306, 2008.

8. R.K. Dhanaraj, S.K. Islam, V. Rajasekar, A cryptographic paradigm to detect and mitigate black hole attack in VANET environments. *Wireless Network*, pp. 1–16, 2022.

9. Martin Brandl, Julius Grabner, Karlheinz Kellner, Franz Seifert, Johann Nicolics, Sabina Grabner, Gerald Grabner, "A low-cost wireless sensor system and its application in dental retainers," *IEEE Sensors Journal*, Vol. 9, pp. 255–262, 2009.

10. M. Kumaresan, M. J. Basha, P. Manikandan, S. Annamalai, R. Sekaran, A. S. Kumar, "Stock Price Prediction Model Using LSTM: A Comparative Study," *2023 3rd Asian Conference on Innovation in Technology (ASIANCON)*, Ravet IN, India, pp. 1–5, 2023. doi: 10.1109/ASIANCON58793.2023.10270708

11. Shih-Sung Lin, Min-Hsiung Hung, Ding-Rong Lai, "Development of a RFID-based missile assembly and test management system," *Chung Cheng Ling Hsueh Pao/Journal of Chung Cheng Institute of Technology*, Vol. 37, pp. 185–195, 2009.

12. M. Ramanan, L. Singh, A. Suresh Kumar, A. Suresh, A. Sampathkumar, V. Jain, N. Bacanin, Secure blockchain enabled Cyber–Physical health systems using ensemble convolution neural network classification. *Computers and Electrical Engineering*, Vol. 101, pp. 108058, 2022.

13. T. Cao, E. Bertino, H. Lei, Security analysis of the SASI protocol, "*IEEE Trans. Dependable and Secure Computing*, Vol. 6, no. 1, pp. 73–77, Jan-Mar 2019.

14. R. Shankar, B.K. Sarojini, H. Mehraj, A.S. Kumar, R. Neware, A. Singh Bist, Impact of the learning rate and batch size on NOMA system using LSTM-based deep neural network. *The Journal of Defense Modeling and Simulation*, Vol. 20, no. 2, pp. 259–268, 2023. doi:10.1177/15485129211049782

15. Suresh Kumar Arumugam, Amin Salih Mohammed, Kalpana Nagarajan, Kanagachidambaresan Ramasubramanian, S. B. Goyal, et al., "A novel energy efficient threshold based algorithm for wireless body sensor network", *Energies*, Vol. 15, no. 16, 6095.

16. M.A. Kumar, A.S. Kumar, A Body Area Network Approach for Stroke-Related Disease Diagnosis Using Artificial Intelligence with Deep Learning Techniques. In: Singh, M., Tyagi, V., Gupta, P.K., Flusser, J., Ören, T. (eds) *Advances in Computing and Data Sciences. ICACDS 2022. Communications in Computer and Information Science*, Vol. 1613. Springer, Cham, 2022. https://doi.org/10.1007/978-3-031-12638-3_21

17. A. Mehbodniya, A. Suresh Kumar, K.P. Rane, K.K. Bhatia, B.K. Singh, "Smartphone-based m health and Internet of Things for diabetes control and self-management," *Journal of Healthcare Engineering*, Vol. 2021, 2021. *Engineering*, Vol. 23, pp. 3–13, 2009.

18. Anonymous, "RFID based paper roll management system," *International Paper Board Industry*,Vol. 51, pp. 20–24, 2008.

19. Karen Conneely, "Managing corporate assets with RFID," *Assembly Automation*, Vol 29, pp. 112–114, 2009.

Chapter 5

Blockchain-enabled intelligent Internet of Things system for secured healthcare applications

Siddhant Thapliyal, Mohammad Wazid, and Devesh Pratap Singh
Department of Computer Science and Engineering, Graphic Era
(Deemed to be University), Dehradun, Uttarakhand, India

5.1 INTRODUCTION

The healthcare system is concerned with providing healthcare services and facilities, preserving and enhancing health, guarding against illness. Further concerns are the reduction in the cost of treating illness, and ensuring a high standard of living for everyone. With the development of technology and the growing desire for better medical services among the target population, healthcare systems proceed for changes all over the globe. Every linked object can communicate with each other and with the outside world via the Internet of Things (IoT), which can be treated as the smart communication environment. IoT applications can comprehend maximum capacity thanks to intelligent IoT. Faster access to more precise data is made possible by the included mechanisms of machine learning and artificial intelligence (AI). The intelligent IoT (IIoT) has various applications, i.e., smart healthcare, smart infrastructure, smart farming, smart transportation, smart homes, and many more. The Internet of Medical Things (IoMT) is a practical application of IIoT in the healthcare industry that offers the chance to improve the current healthcare system and give patients quality treatment and a higher quality of life. It is a collection of healthcare systems (medical equipment, software, and other services) that allow for the safe transfer of health-related data between the smart devices. This enables doctors, healthcare providers, and medical testing facilities, which are located far away, to store and share health information digitally. Through Internet-enabled smart devices like smart phones, wearable and implantable medical devices, electronic medical reports (EMR), IoMT offers real-time medical services and help. Reduced healthcare expenses, prompt medical attention, speedy decisions, and improved patient treatment are some further advantages of IoMT. In IIoT, there are billions of connected devices for a variety of applications, i.e., industries, healthcare. Users' privacy and security are becoming the most difficult issues in IIoT due to its exponential increase, and hence should be considered seriously. Unauthorised access to significant patient sensitive data, such as medical information that are used to make life-altering

DOI: 10.1201/9781003362685-5

choices, is one of several security and privacy breaches that could happen in the healthcare systems. In order to endanger the lives of patients, further hostile behaviours include altering health data, stealing medical devices, breaching into hospital networks and exploiting shared and stored information. In order to combat the threats and attacks against IoMT, it is essential to investigate some optimised security solutions. Blockchain, one of many security mechanisms, has the power to tackle the limitations of conventional approaches to deal with users' privacy and security, and it is thought to be the foundation of secure future IIoT applications (i.e., smart healthcare) because of its many advantages, including improved speed, increased efficiency, true traceability, and increased security. Consequently, combining blockchain with IIoT can increase resistance to various assaults, resulting in better healthcare services delivered in a secure and timely manner. A blockchain is a distributed database, which is shared among various participants in a peer-to-peer (P2P) network. It contains records of every digital event or transaction that has ever taken place. The authenticity of the distributed ledger of transactions (DLT) in blockchain has been determined by peers on a decentralised network protected by cryptography algorithms. In DLT, records are maintained in the form of immutable blocks, which are linked to one another to form a chain.

Blockchain fosters trust because of the following characteristics:

- **Programmable:** Each blockchain can be maintained using some programming language, i.e., it can be created in the form of a smart contract via the solidity programming language.
- **Secure:** Due to the implemented hashing mechanism, the data that is kept in the blockchain is difficult to alter or modify. Additionally, since all data (transactions) are maintained in encrypted form, information leakage is also unfeasible.
- **Anonymous:** Since the identities of the entities are preserved, an attacker cannot determine who is speaking with whom.
- **Distributed:** Blockchain is accomplished via a "peer-to-peer distributed network" as a shared distributed ledger by all authorised mining nodes.
- **Time-stamped:** Every block in a blockchain contains a timestamp values to guarantee its freshness and to make it easier for organisations to pinpoint the moment a particular record was saved in the block.
- **Immutable:** The blockchain constantly stores data in the form of blocks, making it nearly impossible for an attacker to alter a block. If attempts to update or modify any block, they must do it by updating or altering a larger number of blocks than the predefined threshold value, which is almost unfeasible.
- **Unanimous:** The network's miner nodes use the consensus algorithm to add a new block to the blockchain. According to the consensus mechanism, a majority of miner nodes (i.e., 75%) must agree on its addition

before a block can be added to the current blockchain, so the network's nodes work together to complete the task.

Based on its deployment and characteristics, blockchain is divided into different categories. Their details are given below.

- **Public blockchain:** A "permission-less blockchain" is another name for a public blockchain. Everyone can participate in this blockchain and be a part of it by running a node, mining a block, or carrying out transactions. A couple of instances of public blockchains include Bitcoin and Litecoin.
- **Private blockchain:** "Permissioned blockchain" is another name for a private blockchain. Participation is limited in this case because only chosen ones or members of an organisation are allowed to access the blockchain. Examples of private blockchain include the Multichain and Hyperledger projects (i.e., Fabric, Sawtooth), as well as a personal health records system's blockchain.
- **Consortium blockchain:** It is said to be "partially-decentralised or semi-decentralised." Instead of just one organisation like in a private blockchain, it is governed by a number of organisations. The member organisations have the right to take participation via mining, acting as a complete node, etc. Samples of consortium blockchain include "R3 and the Energy Web Foundation (EWF)."
- **Hybrid blockchain:** It combines the features of "public and private blockchain systems."

5.1.1 Chapter contributions

The following are the important contributions of the chapter:

- We discuss various threats and attacks of blockchain-enabled intelligent IoT architecture for secured healthcare system.
- We describe the architecture of blockchain-enabled intelligent Internet of Things system for secured healthcare applications (BIoTS-SH).
- We present an attack model to explore potential BIoTS-SH threats.
- We also give a comparison and analysis of different blockchain-based mechanisms for a safe and secure healthcare system, in which we analyse the various "security and functioning elements" of various schemes.

5.1.2 Chapter organisation

The rest of the chapter is divided into the following sections. The applications of BIoTS-SH are covered in Section 5.2. The BIoTS-SH security and privacy criteria are highlighted in Section 5.3. Further information on BIoTS-SH security threats and attacks is provided in this section. Then, in Section 5.4, the BIoTS-SH architecture is displayed together with the relevant attack model. An overview of several blockchain-enabled intelligent IoT for healthcare systems schemes is given in Section 5.5. A comparison of several blockchain-enabled intelligent IoT for healthcare systems schemes is provided in Section 5.6. The final section of the chapter is 5.7.

5.2 BIOTS-SH APPLICATIONS

Applications of a blockchain-enabled intelligent Internet of Things system for secured healthcare applications (BIoTS-SH) are numerous. Some of the potential applications are given below [1, 2, 3, 4, 5].

- **Management of day-today operations of healthcare:** A doctor always finds, it hard to examine multiple patients at once. It becomes very problematic in cases of populous countries, such as China and India. BIoTS-SH also allows for a prompt reaction to patients from concerned medical professionals, i.e., any available doctor can respond. This means that activities may be handled as needed. As a result, BIoTS-SH aids in the entire administration of hospital operations [6].
- **Disease detection and prevention:** BIoTS-SH helps in the early diagnosis of different illnesses, it can be asthma attack, heart attack, cancer, or anything. Smart healthcare equipment (for example, wearable healthcare devices) continually observes patients' health and gives notifications in the event of a health emergency. Smart healthcare equipment, for example, monitor the patient's real-time results of blood pressure level, oxygen level, or heart rate and send notifications to the appropriate healthcare workers in the event of a heart failure crisis. As a result, BIoTS-SH is useful for illness identification and prevention [2].
- **Management of the drug supply chain:** Illegal pharmaceutical counterfeiting can occur on drug shipments in transit from the manufacturing facility to the patient's delivery. The blockchain-based method of BIoTS-SH can be used to control the safe medicine supply chain [7, 8, 9]. The aforementioned system allows for the placement of "smart tags" on medication bags, which supports effective medication supervision and delivery. This offers safeguards against "anti-counterfeiting of medications". In other cases, "radio frequency identification (RFID)" tags can be utilized to safeguard drug packs against counterfeit medication. A blockchain may be used to automate the entire process, starts with the production to their delivery to the customer. If hostile actors attempt to do counterfeiting of a bag, the deployed "blockchain-based anti-counterfeiting technology" can identify such an incidence. In such a case, the patient receives the original drug, which helps in the better treatment of their medical issues.

- **Urgent care facility:** Urgent care is also available at BIoTS-SH. The majority of BIoTS-SH users and patients communicate with one another over the Internet. As a result, following patients' health situations is a breeze. If doctors are not accessible at a given time or location, other doctors who are available can be consulted. Furthermore, if healthcare equipment and drugs are not readily available, this may be swiftly arranged since BIoTS-SH keeps track of all of these things. As a result, in the event of an emergency, a patient can receive urgent care [10, 11].
- **Data portal for healthcare:** An information portal of healthcare data is a very essential component of BIoTS-SH. Access to healthcare data to the corresponding patients can be provided. It can be checked through any device, like a tablet, laptop, desktop, or a smartphone. It covers all of the information kept in the form of personal health records, such as the patient's disease, continuing treatment, medical consultation, immunisation, and so on. Other services may include appointment scheduling, bill viewing, health insurance data, and online bill paying. In certain circumstances, portals allow patients to communicate with healthcare professionals. Now it's not required to wait for the booking of appointments for the long hours. Instead of that, users can log into the portal and can see the availability of doctors there. After that they can consult accordingly. Additionally, the booking of medications and lab testing can also be done [11].

5.3 BIOTS-SH SECURITY AND PRIVACY

As previously said, BIoTS-SH has a substantial impact on people's lives due to its applicability in a wide range of applications. However, it has a variety of security and privacy issues. Below are the specifications of BIoTS-SH security requirements, threats, and attacks [12].

5.3.1 Security requirements in BIoTS-SH

The following are the security requirements of BIoTS-SH.

- **Health record confidentiality:** BIoTS-SH utilises delicate patient medical services information. Therefore, keeping patients' very own wellbeing records hidden becomes basic. Wellbeing records can take any structure, for example, who has which infirmity, which treatment is being utilised, which clinical insurance contracts are being utilised, etc. To keep this data hidden, a few information encryption methods (like symmetric key or public key) can be used.

These encryption calculations guarantee the privacy of both put away information and information on the way [13].

- **Health record integrity:** Unapproved redesigns ought to be stayed away from on BIoTS-SH devices. To keep up with the respectability of medical care records, hash calculations, for example, SHA256 ought to be utilised. Utilising this strategy, a hash esteem is produced at the shipper's end and afterward attached to the first message. One more hash esteem is determined and contrasted with the got hash esteem at the recipient's end. Assuming the hash values are the same, it is viewed as that the got message is the first. In any case, the message in the channel was altered. It is likewise suggested that any delicate certifications, for example, passwords, be kept as hash values in the servers' data sets, and that when a client attempts to sign into the framework, another hash esteem be made, which ought to be checked with the secret word's put away hash esteem. Assuming that matches, the client ought to be respected authentic and permitted to sign in [14, 15].
- **The system's and its resources' availability:** All BIoTS-SH resources, for instance, various smart devices, servers and data resources should be within easy reach to legitimate parties for 24/7 with no denial of service. Meanwhile, we must protect the BIoTS-SH infrastructure from "denial of service (DoS) and distributed denial of service (DDoS) attempts." Because when they happen, the system and the associated devices and data resources becomes unavailable to the genuine users. As a result, it is vital to incorporate some form of "intrusion detection and prevention" techniques to mitigate these attacks and makes the required resources available to authorised people [16, 17].
- **The recentness of healthcare data:** The fresh form of data is always beneficial in making the right decisions at the right time, especially in the healthcare area. Thus there should be some certainty that the relevant authorities will get new data. Some form of timestamp values provided to the transmitted messages [10, 12, 18].

5.3.2 Security threats and attacks in BIoTS-SH

BIoTS-SH may be subjected to a variety of passive and aggressive assaults. Some BIoTS-SH possible threats are addressed below [12].

- **Replaying of information attack:** The adversary attempts to replay past messages in this hostile conduct. This means that a network object gets previous messages rather than new ones. In traded communications, freshly created timestamp values might be employed for security purposes. At the receiver's end, these timestamp values are likewise checked. If the message is properly verified, then the receiver accepts it. If not, then it will be rejected.
- **Man-in-the-middle (MiTM) attack:** In the malicious MiTM task, it attempts to update the previously transmitted messages and then transmits

the current message to the different entity present in the network. To provide security, nonce values are being used, which are generated randomly in two-way communications, coupled with confidential keys and confidential numbers that are unknown to A. As a result, in such a deployment, A cannot properly edit the exchanged messages.

- **Impersonation attack:** In this criminal conduct, attempts to produce messages in place of a lawful network device/user, then sends the forged message to the addressee in order to include him/her in the conversation. That party believes he or she is communicating with the original party, but in reality, he or she is communicating with an intruder. To provide security, nonce values must be used, which should be generated randomly in exchanged communications, coupled with secret keys and secret numbers that are unknown to an intruder. As a result, in such a deployment, an intruder cannot produce original messages in place of an authorised network entity.

- **Privileged insider attack:** In the malicious task of a privileged insider attack, the trusted user of registration authority becomes the attacker. This user has access to the secret registration information of the different devices and users. Then he or she attempts to derive the secret credentials belonging to genuine network organisations. Following the execution of these malicious operations, the privileged insider user attempts various related attacks such as MiTM, impersonation, and unauthorised session key calculation. To ensure security, it is suggested that all registration information should be deleted from the trusted registration authority's database.

- **Unauthorised session key computation attack:** In this malicious attempt, tries to execute an unauthorised session key computation attack on a deployed security technique. It is recommended to use "short term secrets (i.e., timestamp values, random nonce values) and long term secrets (i.e., secret keys and identities)" for the estimation of session keys. With such security features in place, he/she is unable to execute an unauthorised session key computation attack. As he/she does not have information of "short term secrets and long term secrets."

- **Spreading attack and malware injection:** Malware attacks of various forms, such as keyloggers, spyware, viruses, worms, trojan horses, ransomware, and so on, can be launched in the BIoTS-SH. This act puts the system in a situation to create a halt and preventing it from providing its services to the genuine users. Some of them even revealed vital healthcare data to unauthorised third parties. Malware assaults are often launched by a "bot (Internet attacker system) or the botnet (network of attacker system)". A sophisticated botnet may bring the system down within few seconds. It is advised to utilise firewalls in conjunction with sophisticated intrusion detection and prevention systems for protection [19, 20].

- **Scripting and malicious injection attacks:** SQL injection, cross-site scripting, and cross-site request forgery are malicious kinds of hacking attempts, which can also be performed on a healthcare system. Under these attacks, the secret data, like, personal health records, passwords, credit card numbers, etc., can be revealed to the malicious actors (i.e., hackers). Furthermore, they may also try to change some of the health-related information. It is advised that firewalls (i.e., web application firewall) be used in conjunction with sophisticated prevention systems and intrusion detection for protection.
- **Denial of service/distributed denial of service attacks:** BIoTS-SH systems and its associated resources become unavailable to the original users under the influence of a "denial of service (DoS) or distributed denial of service (DDoS) attack." These attacks may be carried out via malicious jobs such as "SYN flood, HTTP flood, UDP flood, TCP flood, and so on". It is advised that firewalls be used in conjunction with sophisticated "intrusion detection and prevention systems for protection."
- **Consensus attacks:** To breach the security of BIoTS-SH, can launch several consensus assaults such as "the 51% attack, the long-range attack", the balancing attack, and the sybil attack. Under these malicious attempts, it tries to create or put a fake blockchain into the system or in other cases he/she tries to change a fraction of blocks of the blockchain. The solutions like "deployment of a huge number of nodes and high network hashing power are two possible options."

5.4 USED SYSTEM MODELS IN BIOTS-SH

In this section, we discuss the architecture of a blockchain-enabled intelligent Internet of Things system for secured healthcare applications (BIoTS-SH). Apart from that the attack model of BIoTS-SH is also provided [11].

5.4.1 Architecture of BIoTS-SH

The architecture of a blockchain-enabled intelligent Internet of Things system for secured healthcare applications is given in Figure 5.1. There is a patient who has some implantable/ wearable smart healthcare devices, i.e., a heart pacemaker. Smart healthcare devices keep track of patients' health and securely transmits information about that health to a nearby personal server. "Personal digital assistants (PDAs)" are devices that can be deployed and can be treated like a personal server. From the obtained healthcare data, the personal server next generates the partial blocks and sends them to the appropriate cloud server via the access point (i.e., gateway node). Each partial block contains a number of encrypted transactions that represent patient health information. All transactions are encrypted using the personal server's public key. The peer-to-peer cloud server (P2PCS)

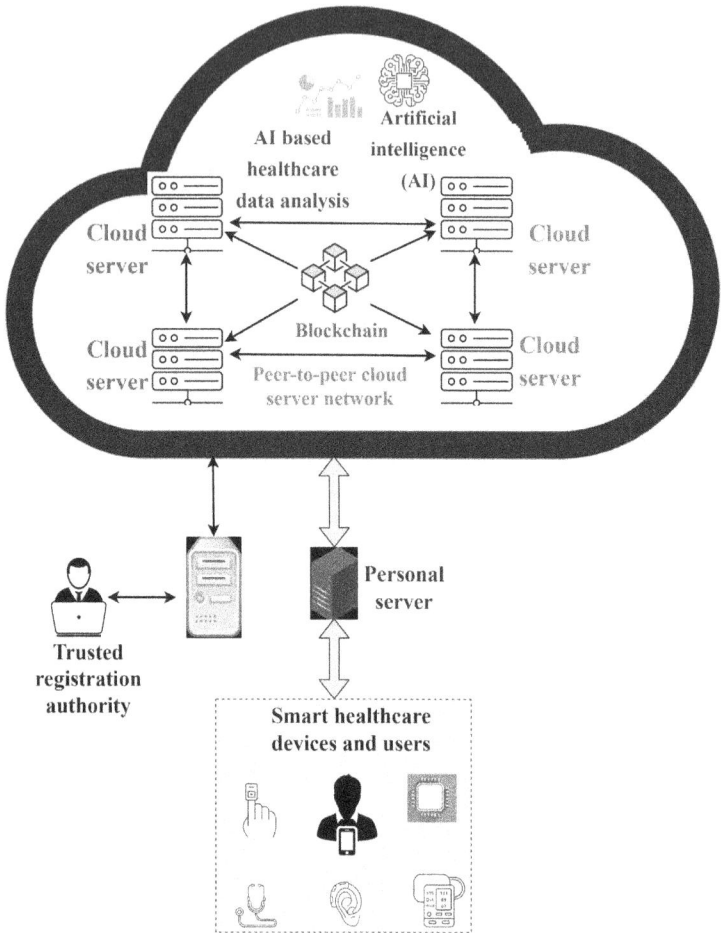

Figure 5.1   Architecture of a blockchain-enabled intelligent Internet of Things system for secured healthcare applications.

network will then receive the entire block that the cloud server created from the partial block that it had previously received and verify it before adding it to the blockchain. A leader in the P2PCS network will be chosen from the current cloud servers. The consensus procedure for the block's insertion to the blockchain and verification will then be initiated by that leader. A blockchain will be added to the blockchain if the leader receives the necessary number of commits from the followers' nodes (other cloud servers). A distributed ledger is used to maintain blockchain, and it is accessible to all miners. The system's other miner nodes will be updated when a new block is added to the blockchain. The sensitive nature of the healthcare information means that it should never be altered or released.

Blockchain technology is therefore very beneficial to achieving these information security objectives. All system users and devices must be registered with a trusted registration authority, which serves this function. Some users, such as doctors, nurses, and laboratory staff, are also interested in obtaining the healthcare information of the patients. They can utilise this information to write prescriptions, perform medical procedures, analyse lab test results, and write reports, among other things. After following the necessary user authentication processes, the authorised and legitimate system users can securely access the data of BIoTS-SH. Later on, some prediction can also be made on the stored healthcare data through some machine learning algorithms (i.e., chances of getting a heart attack, chances of a diabetic shock, etc.). The flow of different activities of BIoTS-SH is given below [13].

- There is the device and user registration by a reliable registration authority. Then credentials will be stored in the memory of various devices.
- Further, there is the execution of "authentication and key establishment" among users, smart healthcare devices, personal server, and cloud servers.
- Transmission of secure healthcare data between smart healthcare devices and personal servers.
- Personal server aggregates the healthcare data. Following that, healthcare data is transformed into transactions. Following that, all transactions will be encrypted using the personal server's public key.
- Additionally, from the encrypted transactions, personal servers generate partial blocks. Each partial block has information for the block's owner, public key of personal server and encrypted transactions. The established session key is then used to send partial blocks securely to the corresponding cloud server.
- By including necessary variables such as the "block's ID (number), timestamp value, merkle tree root, hash of the current block, hash of the previous block, and signature of this block," the cloud server produces the full block from the partial block.
- The P2PCS network's leader initiates and conducts the consensus process utilising a consensus algorithm, i.e., "practical Byzantine fault tolerance (pBFT)." The new block will be included on the blockchain after all miners commit on it. A distributed ledger called the blockchain is kept up to date and is available to miner nodes.
- After completing the necessary stages of the authentication system, users (such as doctors, nurse) who are interested in accessing the data of BIoTS-SH can acquire it.

- Later on, some prediction can also be made on the stored healthcare data through some machine learning algorithms (i.e., chances of getting a heart attack, chances of a diabetic shock, etc.).

5.4.2 Attack model of BIoTS-SH

Here, we talk about the potential threats related to BIoTS-SH. The guidelines of the well-known Dolve-Yao (DY) model can be applied when creating a "blockchain-enabled intelligent IoT architecture for secured healthcare system [21]." According to the DY paradigm, users and communication devices communicate via an unsecure public channel (i.e., Internet). As a result, the active network adversary has the ability to intercept, alter, postpone, or even erase the messages that these entities exchange [12]. Some of the deployed smart healthcare equipment can also be physically taken over by an attacker, who will then attempt to use a sophisticated power analysis assault to steal information from their memory [22]. Additionally, the "Canetti-Krawczyk adversary (CK-adversary) model" is a significant model that can be used in the design of a "Blockchain-enabled intelligent IoT architecture for secured healthcare system [23]". The CK-adversary model has all of the same capabilities as the DY model, and additionally can compromise "session states (i.e., session keys)," which are established between the communicating entities for secure communication. As it handles the registration of numerous network users and devices, trusted authority is regarded as the network's sole trusted authority. The cloud servers and personal servers that are used to implement blockchain technology are regarded as the network's semi-trusted entities. In order to guard against physical device compromise attacks, it is also anticipated that personal servers be kept inside "physical locking system."

5.5 BLOCKCHAIN-ENABLED PERSONAL HEALTH RECORD SECURITY SOLUTIONS

Authentication and access control techniques are regarded as two fundamental security services used in a variety of applications [24, 25, 26, 27, 28]. In this part, we talk about the security methods that may be used for "blockchain-enabled intelligent IoT architecture for secured healthcare system [20]." Following that, we compare their security and functionality qualities. The comparison research is important in determining which scheme is more secure and offers more functionality characteristics.

5.5.1 Examination of the approach by Younis et al. [29]

- They presented a blockchain-based architecture for telehealth applications. They used "blockchain-enabled decentralized access control and logging mechanism."

- To safeguard the delivery of patient's information, they offered a data-driven protection mechanism.
- They verified the security of their scheme through AVISPA.
- They have tested how well their strategy performed in terms of important performance parameters as well as computation time.

5.5.2 Examination of the approach by Saha et al. [30]

BIoTS-SH may be utilised with a "blockchain-based access control system for IoT-enabled healthcare," according to Saha et al. [30]. The following is a brief overview of Saha et al.'s scheme:

- Access control was addressed via a private blockchain-based approach.
- Their system proved beneficial to the trusted group of hospitals in terms of preserving and securely exchanging of healthcare data.
- They had used "ECC-based signature technique." Their scheme's security was based on solving the "elliptic curve discrete logarithm problem (ECDLP)" and "collision-resistant one-way hash function."
- Their plan was divided into sections, such as "registration phase, login and access control phase and blockchain construction phase."
- They evaluated the scheme's security against "replay attacks, MiTM attacks, impersonation attacks, and ephemeral secret leakage (ESL) assaults."

5.5.3 Examination of the approach by Xiang et al. [9]

Subsequently, for the purpose of retaining health information, Xiang et al. suggested new permissioned blockchain-envisioned identity management and user authentication strategy. The following list includes typical attributes of Xiang et al.'s approach:

- In their scheme a "permissioned blockchain-enabled identity management and user authentication mechanism" was used for maintaining the healthcare data.
- It has important steps like, "installation phase, enrollment phase, login phase, authentication and key agreement phase, and password update phase." However, other critical aspects, such as "dynamic device addition and key revocation," was absent from their approach.
- To determine its defence against a range of potential attackers, a security assessment was also carried out.

5.5.4 Examination of the approach by Neha et al. [11]

Neha et al. [11] developed a "blockchain-enabled authenticated key management technique for IoMT deployment," which is also applicable to BIoTS-SH. The overview of Neha et al.'s scheme is given below:

- They considered private blockchain when developing their plan.
- In the design, the quick "one-way cryptographic hash function" and "bitwise XOR operations" were used.
- They have also provided the formal security verification of their protocol via "AVISPA (Automated Validation of Internet Security Protocols and Applications)" tool to demonstrate its resistance to replay and MiTM attacks. In addition, an informal security study was performed to demonstrate its protection against several potential threats.
- Additionally, the implementation was supplied to them so that they could gauge its impact on performance indicators like computation speed and transactions per second.

5.5.5 Examination of the approach by Xu et al. [8]

Xu et al. presented a "blockchain-enabled smart healthcare system for large-scale health data privacy." The following are typical aspects of Xu et al.'s system:

- They proposed a "blockchain-enabled smart healthcare system for safe healthcare data storage and interchange."
- The provided model distinguished between data publication transactions and all other potential access control operations. The "interplanetary file system (IPFS)" might be used to encrypt and store health records.
- There was also a key revocation feature, which allowed other parties, such as a doctor, to have their keys revocation.

5.5.6 Examination of the approach by Islam and Shin [31]

Islam and Shin presented a "blockchain-enabled approach for healthcare data security [31]." Some of the most fundamental aspects of Islam and Shin's system are as follows:

- A blockchain-based scheme for the safe gathering and exchange of healthcare data using unmanned aerial vehicles (UAVs) was presented.
- They provided a "2-phase authentication system." A threat model was also created to encompass the numerous attacks of such a communication.
- A security study was also undertaken to demonstrate its resistance to several possible attacks.
- The ethereum platform was used to build the consortium blockchain on it. Throughput, latency, and block size were among the network performance parameters, which were computed and analysed.

5.6 COMPARATIVE ANALYSIS

In this section, a comparative study on blockchain-enabled security scheme for healthcare system is conducted. The different scheme, for instance, Neha et al.'s scheme [11], Saha et al.'s scheme [30], Xiang et al.'s scheme [9], Xu et al.'s scheme [8], Islam and Shin's scheme [31], and Younis et al. 's scheme [29] are analysed and compared. The details of comparisons are given in Table 5.1.

During the comparative study, we considered the following important security and functionality features:

- SFF_1: "provides mutual authentication/access control"
- SFF_2: "supports anonymity property"
- SFF_3: "supports untraceability property"
- SFF_4: "provides session-key agreement"
- SFF_5: "provides session key security under CK adversary model"
- SFF_6: "provides data confidentiality"
- SFF_7: "provides data integrity"

Table 5.1 Comparison of security and functional attributes

Feature	Neha et al. [11]	Saha et al. [30]	Xiang et al. [9]	Xu et al. [8]	Islam and Shin [31]	Younis et al. [29]
SFF_1	✓	✓	✓	✓	✓	✓
SFF_2	✓	✓	✓	✓	✓	×
SFF_3	✓	✓	✓	✓	✓	×
SFF_4	✓	✓	✓	×	×	×
SFF_5	✓	✓	×	×	×	×
SFF_6	✓	✓	✓	✓	✓	✓
SFF_7	✓	✓	✓	✓	✓	✓
SFF_8	✓	✓	✓	✓	✓	✓
SFF_9	✓	✓	✓	✓		✓
SFF_{10}	✓	NA	✓	NA	NA	NA
SFF_{11}	✓	NA	✓	NA	NA	NA
SFF_{12}	✓	NA	×	NA	NA	NA
SFF_{13}	✓	×	×	×	×	×
SFF_{14}	✓	×	×	×	×	×
SFF_{15}	✓	NA	✓	NA	NA	×
SFF_{16}	✓	✓	✓	✓	✓	✓
SFF_{17}	✓	×	✓	×	×	✓
SFF_{18}	×	×	×	×	×	×
SFF_{19}	✓	×	✓	✓	✓	×
SFF_{20}	✓	✓	✓	✓	✓	✓

Note: ×:"a scheme is insecure against that particular attack or does not support a specific feature"; ✓:"a scheme is secured against that particular attack or supports a specific feature"; NA:"not applicable in a scheme".

- SFF_8: "protection against strong replay attack"
- SFF_9: "protection against man-in-the-middle attack"
- SFF_{10}: "availability of efficient login phase"
- SFF_{11}: "availability of password update phase"
- SFF_{12}: "availability of biometric update phase"
- SFF_{13}: "availability of dynamic controller node (personal server) addition phase"
- SFF_{14}: "availability of dynamic smart healthcare device addition"
- SFF_{15}: "protection against stolen mobile device/programmer attack"
- SFF_{16}: "protection against impersonation attack"
- SFF_{17}: "provides formal security verification using AVISPA/SCYTHER tool"
- SFF_{18}: "provides formal security analysis under real-or-random (RoR) model"
- SFF_{19}: "provides practical implementation"
- SFF_{20}: "blockchain enabled security"

From Table 5.1, It is obvious that the plans of Saha et al. [30], Xiang et al. [9], Xu et al. [8], Islam and Shin [31], and Younis et al. [29] do not provide required security and functionality features like, "provides session-key agreement", "provides session key security under CK adversary model", "availability of password update phase", "availability of biometric update phase", "availability of dynamic controller node (personal server) addition phase", "availability of dynamic smart healthcare device addition", "provides formal security verification using AVISPA/SCYTHER tool" and "provides formal security analysis under real-or-random (RoR) model". However, the scheme of Neha et al. [11] achieved most of the required "security and functionality features". Therefore, Neha et al.'s scheme [11] seems suitable for the secure transmission and storage of healthcare data of "blockchain-enabled intelligent IoT architecture for secured healthcare system."

5.7 CONCLUSION

Blockchain innovation is exceptionally useful for secure transmission and capacity of medical care information. As we probably are aware, data arrangement of medical care is worked through different structures, i.e., e-health. All these systems need to maintain healthcare data. We must securely handle and store healthcare data on an insecure medium. Blockchain technology is used in these systems to protect healthcare data from numerous forms of potential attacks. "Blockchain-enabled intelligent IoT architecture for secured healthcare system (BIoTS-SH)" was the topic of our discussion. We also provided a suitable attack model for BIoTS-SH. A comparison of the most advanced blockchain-enabled security protocols for healthcare data is given, which seems suitable for the selection of a secure scheme for BIoTS-SH.

REFERENCES

1. M. Asif-Ur-Rahman, F. Afsana, M. Mahmud, M. S. Kaiser, M. R. Ahmed, O. Kaiwartya, and A. James-Taylor. Toward a Heterogeneous Mist, Fog, and Cloud-Based Framework for the Internet of Healthcare Things. *IEEE Internet of Things Journal*, 6(3):4049–4062, 2019.
2. R. Chowdhury. IoT in Healthcare: 20 Examples That'll Make You Feel Better. www.ubuntupit.com/ iot-in-healthcare-20-examples-thatll-make-you-feel-better/ Accessed on March 2020.
3. N. Garg, M. Wazid, A. K. Das, D. P. Singh, J. J. P. C. Rodrigues, and Y. Park. BAKMP–IoMT: Design of Blockchain Enabled Authenticated Key Management Protocol for Internet of Medical Things Deployment. *IEEE Access*, 8:95956–95977, 2020.
4. A. Islam, and S. Young Shin. A Blockchain-Based Secure Healthcare Scheme with the Assis–tance of Unmanned Aerial Vehicle in Internet of Things. *Computers & Electrical Engineering*, 84:106627, 2020.
5. A. Sureshkumar, R. Samson Ravindran. Swarm and Fuzzy Based Cooperative Caching Framework to Optimize Energy Consumption Over Multimedia Wireless Sensor Networks. *Wireless Personal Communications*, 90, 961–984, 2016. https://doi.org/10.1007/s11277-016-3274-0
6. V. Bhuse, and A. Gupta. Anomaly Intrusion Detection in Wireless Sensor Networks. *Journal of High Speed Networks*, 15(1):33–51, 2006.
7. A. K. Das, S. Zeadally, and M. Wazid. Lightweight Authentication Protocols for Wearable Devices. *Computers & Electrical Engineering*, 63:196–208, 2017.
8. A. Suresh Kumar, S. Jerald Nirmal Kumar, Subhash Chandra Gupta, Anurag Shrivastava, Keshav Kumar, and Rituraj Jain, et al., "IoT Communication for Grid-Tie Matrix Converter with Power Factor Control Using the Adaptive Fuzzy Sliding (AFS) Method", *Scientific Programming*, 03:1–11, 2022.
9. V. Odelu, A. K. Das, and A. Goswami. SEAP: Secure and Efficient Authentication Protocol for NFC Applications using Pseudonyms. *IEEE Transactions on Consumer Electronics*, 62(1):30–38, 2016.
10. A. K. Das, M. Wazid, N. Kumar, M. K. Khan, K. R. Choo, and Y. Park. Design of Secure and Lightweight Authentication Protocol for Wearable Devices Environment. *IEEE Journal of Biomedical and Health Informatics*, 22(4):1310–1322, 2018.
11. J. Fruhlinger. Ransomware Explained: How it Works and How to Remove it, 2018. www.csoonline.com/article/3236183/what-is-ransomware-how-it-works-and- how-to-remove-it.html Accessed on March 2020
12. P. Nasrullah. Internet of Things in Healthcare: Applications, Benefits, and Chal–lenges. www.peerbits.com/blog/internet-of-things-healthcare-applications-benefits- and-challenges.html Accessed on March 2020
13. T. S. Messerges, E. A. Dabbish, and R. H. Sloan. Examining Smart-Card Security under the Threat of Power Analysis Attacks. *IEEE Transactions on Computers*, 51(5):541–552, 2002.

14. S. Challa, M. Wazid, A. K. Das, and M. K. Khan. Authentication Protocols for Implantable Medical Devices: Taxonomy, Analysis and Future Directions. *IEEE Consumer Electronics Magazine*, 7(1):57–65, 2018.
15. S. Challa, M. Wazid, A. K. Das, N. Kumar, A. G. Reddy, E. Yoon, and K. Yoo. Secure Signature-Based Authenticated Key Establishment Scheme for Future IoT Applications. *IEEE Access*, 5:3028–3043, 2017.
16. Suresh Kumar Arumugam, , Amin Salih Mohammed, Kalpana Nagarajan, Kanagachidambaresan Ramasubramanian, and S. B. Goyal, et al., "A Novel Energy Efficient Threshold Based Algorithm for Wireless Body Sensor Network", *Energies*, 15(16), 6095.
17. D. Mishra, A. K. Das, and S. Mukhopadhyay. A Secure and Efficient ECC-Based user Anonymity-Preserving Session Initiation Authentication Protocol using Smart Card. *Peer-to–Peer Networking and Applications*, 9(1):171–192, 2016.
18. A. K. Das, M. Wazid, A. R. Yannam, J. J. P. C. Rodrigues, and Y. Park. Provably Secure ECC-Based Device Access Control and Key Agreement Protocol for IoT Environment. *IEEE Access*, 7:55382–55397, 2019.
19. What is Spyware? And How to Remove it, 2020. https://us.norton.com/inter netsecurity-how- to-catch-spyware-before-it-snags-you.html Accessed on March 2020
20. D. Dolev, and A. C. Yao. On the Security of Public Key Protocols. *IEEE Transactions on Information Theory*, 29(2):198–208, 1983.
21. M. Debe, K. Salah, R. Jayaraman, and J. Arshad. Blockchain-Based Verifiable Tracking of Resellable Returned Drugs. *IEEE Access*, 8:205848–205862, 2020.
22. S. Jangirala, A.K. Das, and A. V. Vasilakos. Designing Secure Lightweight Blockchain–Enabled RFID-Based Authentication Protocol for Supply Chains in 5G Mobile Edge Com–puting Environment. *IEEE Transactions on Industrial Informatics*, 16(11):7081–7093, 2020.
23. R. Canetti, and H. Krawczyk. Analysis of Key-Eexchange Protocols and Their use for Building Se–cure Channels. In *International Conference on the Theory and Applications of Cryptographic Techniques–Advances in Cryptology (EUROCRYPT'01)*, 453–474. Springer, Innsbruck (Tyrol), Austria, 2001.
24. M. Ramanan, L. Singh, A. Suresh Kumar, A. Suresh, A. Sampathkumar, V. Jain, and N. Bacanin. Secure Blockchain Enabled Cyber–Physical Health Systems using Ensemble Convolution Neural Network Classification. *Computers and Electrical Engineering*, 101, 108058X, 2022.
25. S. Chatterjee, A. Das, and J. Sing. An Enhanced Access Control Scheme in Wireless Sensor Networks. *Ad-Hoc and Sensor Wireless Networks*, 21:121–149, 2014.
26. P.A. Laplante, and N. Laplante. The Internet of Things in Healthcare: Potential Applications and Challenges. *IT Professional*, 18(3):2–4, 2016.
27. S. Challa, A. K. Das, P. Gope, N. Kumar, F. Wu, and A. V. Vasilakos. Design and Analysis of Authenticated Key Agreement Scheme in Cloud-Assisted Cyber–Physical Systems. *Future Generation Computer Systems*, 108:1267–1286, 2020.

28. A. K. Das, A. K. Sutrala, S. Kumari, V. Odelu, M. Wazid, and X. Li. An Efficient Multi–Gateway-Based Three-Factor User Authentication and Key Agreement Scheme in Hierarchical Wireless Sensor Networks. *Security and Communication Networks*, 9(13):2070–2092, 2016.

29. M. Younis, W. Lalouani, N. Lasla, L. Emokpae, and M. Abdallah. Blockchain-Enabled and Data-Driven Smart Healthcare Solution for Secure and Privacy-Preserving. *Data Access IEEE Systems Journal*, 16(3):3746–3757, 2022.

30. M. Kumaresan, M.J. Basha, P. Manikandan, S. Annamalai, R. Sekaran, and A. S. Kumar. "Stock Price Prediction Model Using LSTM: A Comparative Study," *2023 3rd Asian Conference on Innovation in Technology (ASIANCON)*, Ravet IN, India, 2023, pp. 1–5. doi: 10.1109/ASIANCON58793.2023.10270708

31. H. Habibzadeh, K. Dinesh, O. Rajabi Shishvan, A. Boggio-Dandry, G. Sharma, and T. Soyata. A Survey of Healthcare Internet of Things (HIoT): A Clinical Perspective. *IEEE Internet of Things Journal*, 7(1):53–71, 2020.

Chapter 6

Visual efficient easy interactive (VEEI) understanding of teaching in education

*A. Gayathiri[1], S. Mohanapriya[2], S. Bhuvaneswari[1],
P. Sumitra[1], M. Sathiya[1], and S. Annamalai[3]*
[1]PG and Research Department of Computer Science & Application, Vivekanandha College of Arts and Sciences for Women (Autonomous), Namakkal, Tamilnadu, India
[2]Department of Computer Applications, Sona College of Arts & Science, Salem, Tamilnadu, India
[3]School of Computer Science and Engineering, Jain (Deemed-to-be University), Bengaluru, Karnataka, India

6.1 INTRODUCTION

An educator/teacher/professor who shares knowledge, gives education, and teaches about life is the mainstream key for students. It is the responsibility of an educator to mentor students when it is needed. Today's students need to be mentored in every aspect by an educator or teaching professional. Students are from various areas, including rural, urban, semi-urban, and cities. Students' behavior also varies depending on their areas. It is the responsibility of the teacher/professor to handle all kinds of students. As a social responsibility, it is a teacher's or professor's responsibility to set expectations of cleanliness, encourage students to do work by themselves, to think positively, to implement their thinking, to try to write a schedule, to pay for wards' work, to perform house responsibilities, in addition, they should teach students to behave in public, to be realistic, and to not behave as an actor.

As a teacher it is mandatory to categorise the types of students in learning. Even though there are different types of students based on the teacher's view it may differ from person to person. First, there are initial students who need considerable help to learn. These types of students need continuous motivation, help and apart from teaching, notes need to be given back for learning. The next type of student needs teaching and notes, but they do not need more motivation. Thereafter, students need teaching and notes, and the last type of student just needs simple guidance and no notes.

As a teacher we need to concentrate on all these types of students and come together to participate in all activities. This chapter discusses the tools that can be used in education to make the learning more effective and efficient by ensuring all students participant in an interactive manner. These cooperative and interactive tools allow all types of students to be involved in learning development and importantly enables teachers to watch their

DOI: 10.1201/9781003362685-6

students' involvement in learning. As a technical idea, light fidelity, or LED light, is a novel technology that can be utilised as a communication channel. Wireless fidelity (Wi-Fi) is the most widely used form of communication technology. A Wi-Fi network, defined simply, is a wireless router that provides internet access to a variety of devices throughout a home or workplace. Li-Fi, on other hand, is wireless communication technology that sends information and location between devices using light sources instead of using radio frequency.

6.2 LITERATURE REVIEW

Government [1] school students are taken on an educational tour in Tamilnadu. One such occasion involves sending five teachers, 67 students, three Education Department Officials and Education Minister Mr. Anbil Mahesh Poyyamozhi on a "Class Outside Classroom" excursion to Dubai from November, 10 to November, 13. He continued by saying that this supports extracurricular activities among students and prepares them for picking a college and a job.

The authors of paper [2] stated that the most important factor for learning is the teacher. They explained the various learning tools practiced in Indonesia for effective learning. By using the tools, the students understand the teaching materials well, they can verify the concepts, students' difficulties decrease, and the delivery of teaching becomes very easy. The teaching tools have a positive influence on the educational process. The teaching tools are stretchy and adjustable to students' abilities. The students' absorption of the teaching tools has increased and the visualisation increases the students' attention, stimulating critical thinking. This allows students to always remember what has been taught in comparison to just oral teaching and learning.

The author of [3] explains the different types of teaching methods in India. She insists on attending various field excursions in their relative departments. She also suggests that it helps students to become talented in reading, builds confidence, and increases accessibility. The education system in India is now practicing to go on tours, field trips, extension activities, and industrial visits in higher education for every semester.

The author in this paper [4] explains the Kahoot is an Information Systems Strategy, cultivating student engagement, developing classroom energy, and progressing the overall learning experience of students. It is the tool where students are educated with the help of games. As an alternative from teaching using chalk and talk, this type of use of educational games create a center of attention for the students and diminished diversion of the students. Kahoot also helped students learn more by including the creation and blending of pertinent content into the game-playing policy and Kahoot! Platform. Students are eager to experiment with cutting-edge technologies to support their learning, in large part because of their expertise in using mobile technology and gratitude of the apps and games produced for such devices [5].

In [6], two villages in Gujarat were the first Indian villages to get Li-Fi technology. The villages are Akrund and Navanagar, which belong to the Aravalli district of Gujarat, and were converted into the first smart villages with Li-Fi based internet connectivity. Having this capability from the beginning, The Nav Wireless Technology extended its Li-Fi connectivity into schools, then into hospitals, including post offices, and also extended its service to the government offices in these villages, delivering a faster and safer internet connection, all the way through existing electricity lines. It is used in inner-city areas where the radio spectrum is very jam-packed and used in places where fiber-optic cables or the network is not reachable. By using Li-Fi Wireless Optical Communication, the Gujarat's Fiber Grid Network, using internet connectivity using this fiber optic, starts its connectivity from the Akrund Gramme Panchanyat block of buildings to the Navanagar Primary School. This Hybrid Microwave Li-Fi-enabled LED lights are connected to the buildings' existing power connections in schools, hospitals, and post office rooms. They propose to provide speedy internet access in rural parts of Himachal Pradesh and Uttarakhand.

6.3 METHODOLOGY AND TOOLS

6.3.1 Jamboard

Jamboard was built by Google. Its mechanism is operational as an interactive whiteboard to work in Google Workspace and was in earlier times known as G Suite. Its official announcement was on October 25, 2016. Jamboard is a part of the Google suite and works as a great online interactive whiteboard and also a collaborative tool to use with students in a virtual class room. Here we can check questions for students. Teachers use Jamboard to view students visually portray their learning and thinking. This enhances the classroom participation by viewing rather than hearing. Additionally, it gives teachers real-time information on what their students understand. It is an excellent tool for teamwork in physical, virtual, synchronous, and asynchronous environments [7] as shown in Figure 6.1.

There are six ways to use Jamboard to keep your students engaged in your lesson, the foremost way to use Google Jamboard is to insert gif images and sticky notes into the Jamboard. Here students are engaged to create graffiti walls together as a class. This can be done with a blank Jamboard and shared with the whole class. They then collaborate together to add images, facts, and drawings to show what they already know about a topic. A similar version to the graffiti wall is the wonder wall. To create a wonder wall at the beginning, the staff posts the topic of interest and the students are

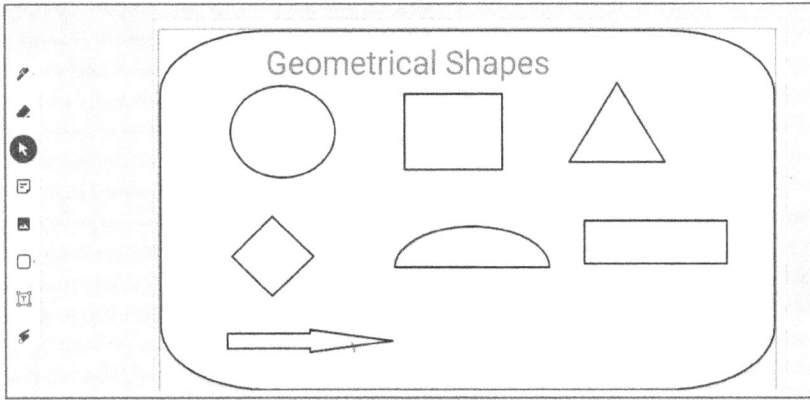

Figure 6.1 A new Jamboard by inserting geometrical shapes using the shape tool bar, the Jamboard also saved with the name, "Geometrical Shapes".

asked to post what they know about the topic interactively. After that is the another interesting feature called Virtual Gallery Walks where a Jamboard can be created and shared with the class and on each one of those Jamboards you can add a different picture or historical document or text that you want students to examine either individually or in breakout rooms where they can go through each of the slides and add their own thoughts and observations, they can then come back as a whole to debrief the gallery work together. The fourth way to use Jamboard is to Create Graphic Organizers, which you can use for mini lessons during your live instruction: there are two ways to do this on Jamboard, the first, is to use a graphic organiser that you can find in the Google image search or one from your own files, the second is to create your own graphic organisers using the shape tools in Jamboard. The fifth way is using Annotations, which is an interactive whiteboard where you are able to highlight, write, draw, and take a screenshot of whatever presentation or worksheet that students see and then to write directly on it and to discuss the article or highlight and add their own annotations. The final way is to Collaborate in all of the Jamboard, i.e., we can combine all of the above ways and if we need to draw even the mathematical diagrams like a Venn diagram, it can also be done and added in the Jamboards [8].

6.3.2 Ziteboard

Ziteboard is a virtual whiteboard collaboration tool used for communication. It is used as effective teaching tool in Skype and on the Zoom classrooms as shown in Figure 6.2. It is a zoomable tool. Mainly it is used as a doodle for all types of students from school to colleges.

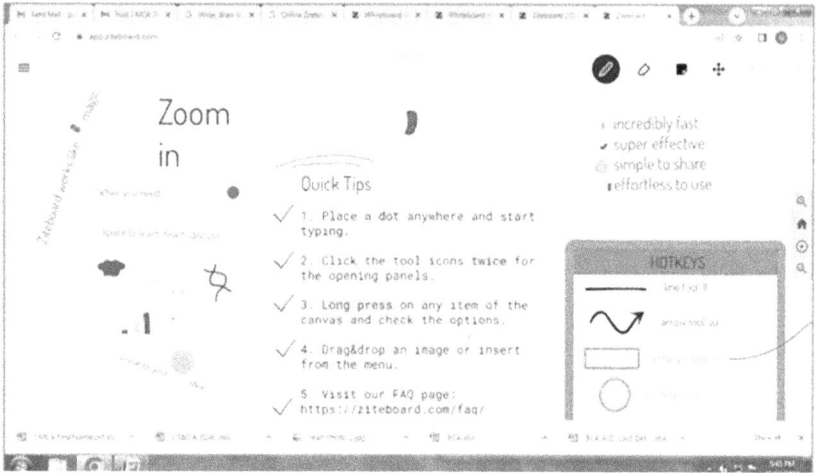

Figure 6.2   Home page screenshot of the Ziteboard application.

This lightweight whiteboard works on any device like laptops, tablets, and mobile devices. For designing things, workflows, and basic whiteboarding information, this Ziteboard is incredibly helpful. It is intended to work in collaboration essentially for the educational institution. It enables freehand drawing, after which we may perform smoothing and shape identification. You can add photos, PDFs, charts, sticky notes, and navigation boxes. Text messaging, as well as audio and video conferencing options, are available for real-time collaboration on interactive whiteboards. High-resolution PDFs can be exported from digital whiteboards [9].

This effective Ziteboard is a very positive tool for teaching, using different types of basic tools. Additionally, we can invite questions during class, which can also be done with the help of the Sticky Notes Tool as shown in Figure 6.3. Here we are able to adjust the font size and color of the Sticky Notes. With the help of the Todo List, we can able to prioritise the tasks, to give the importance of the task. The main advantage of the Ziteboard is we can zoom in and out of the board. We can share the Ziteboard by inviting the students to join the Ziteboard. Through online, we as teachers allow students to work collaboratively and make the session more effective [10].

6.3.3 Lucidchart

Lucidchart's [11] co-founders Ben Dilts and Karl Sun initiated Lucidchart in 2010. They added supplementary products to their collection. At present, many people in the world use Lucidspark on the route to collaborate

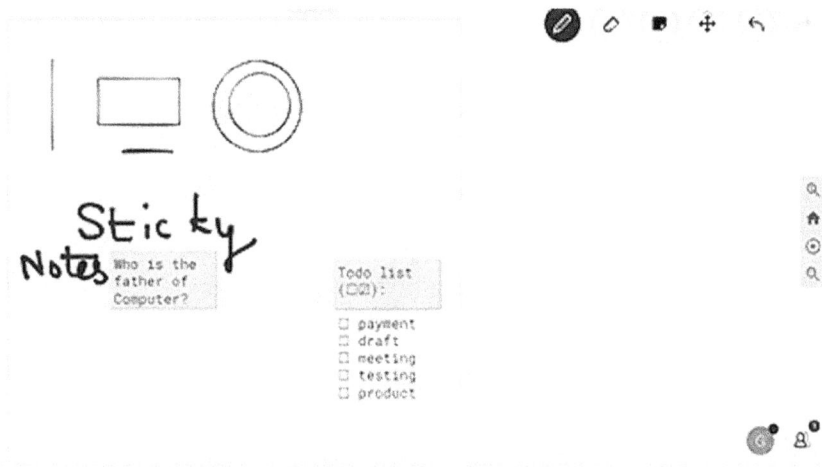

Figure 6.3   Screenshot for adding Sticky Notes, Todo List.

visually. The rationale for using Lucidspark initially is to work in remote interactively with your groups. The next is to facilitate meeting with a digital whiteboard. It allows you to invite your meeting cohort to view the whiteboard at a specific part and allows it to be locked in that place. Lucidspark also has a timer and a make poll vote. In addition to this, we can also create illustrations and a customer path map that depicts your processes or products. Lucidspark furthermore adds one whiteboard where we can make an uncountable number of canvases to hold in one central location. It is also possible to systematise the Sticky Notes, the Note Panel, and modify tags. By developing a SWOT analysis, product roadmap, or customer journey map from within your online whiteboard, you can skip to the finish of important business initiatives. To share ideas with others or to export your board to Lucidchart for further development and execution, invite collaborators as shown in Figures 6.4, 6.5, and 6.6.

6.3.4 Zoom meeting

A cloud-based video conferencing service is Zoom, which is recommended for business and offers features, including webinars, chat, video and audio conferencing and collaboration. During the COVID-19 pandemic, educational institutions used the Zoom application for teaching. The technology was mainly unknown to all teachers who were used to train in the Zoom platform. For two years, Zoom meetings acted as the classroom for students and staff members. It created a remote classroom for us all in the teaching

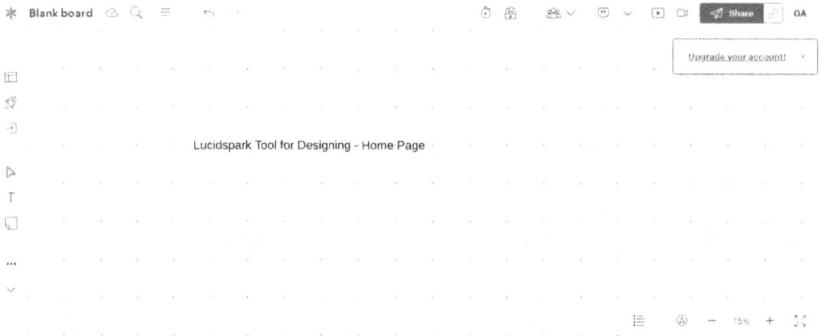

Figure 6.4   The new Blank Board with Pen tool in Lucidspark.

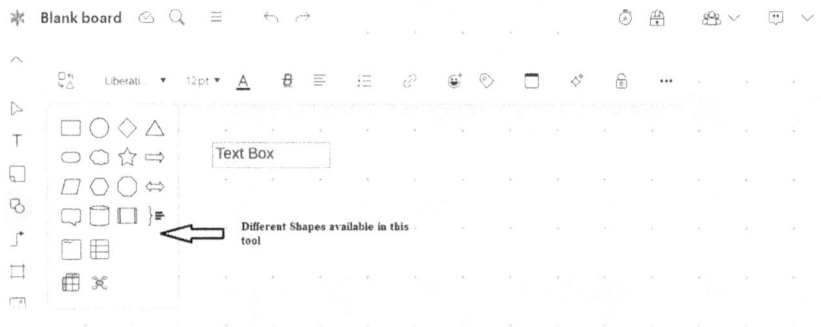

Figure 6.5   A new Blank Board with different shapes in Lucidspark.

community. A Zoom meeting is opened by a normal browser and supported 100 participants at time. When additional subscription is done, it can support 500 participants [12].

The main feature of Zoom meeting is it has a free version, the further features are that Zoom meeting supports a large audience, it is possible to stream a meeting in celebrated Facebook, it supports keeping track of daily events, it is an easy to use environment, and lastly it is scalable [13].

It is easy to set up the Zoom meeting; the organiser can schedule the meeting as an instant meeting or schedule using the calendar. By sharing the Zoom meeting code, this code can be a link or we can enter the code. The invites are able to participate in the meeting using either the browser link or meeting code. Meetings organised are used to control the entry of the participant by allowing them one by one, or by group, or by following the organisation mail-IDs. It does not matter which device, such as smart phones, laptops or systems, are used in the meeting [14] as shown in Figure 6.7.

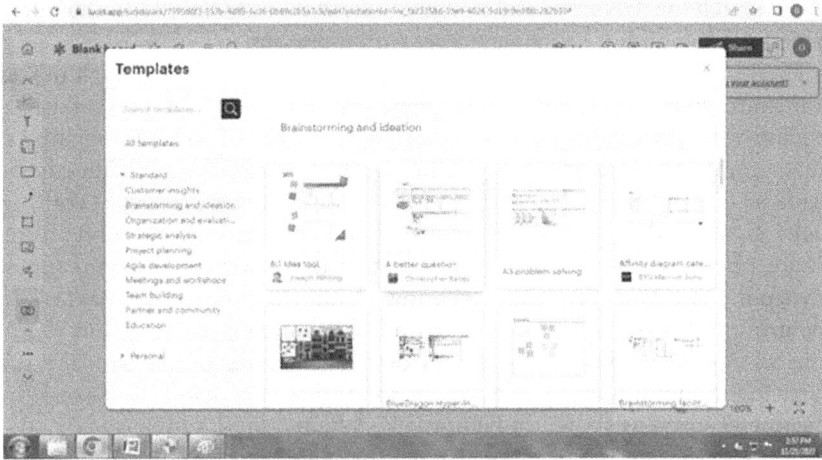

Figure 6.6   The new Blank Board with different templates.

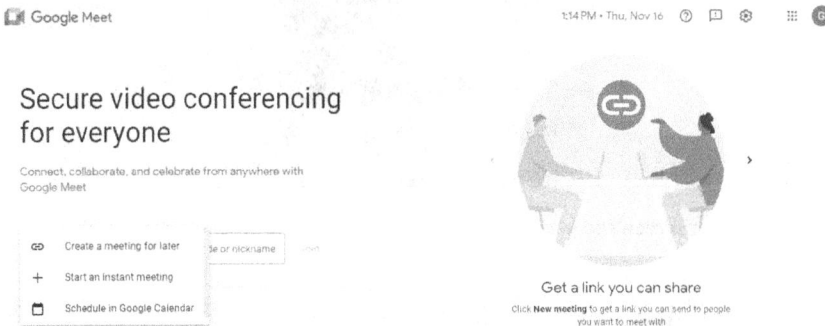

Figure 6.7   The initiation of Google Meet, with three options to create a new meeting.

With high-definition audio and video, people who are seeing and hearing through, have a clear understanding and visualising the video of the particular content. It is easy to configure and conference settings by flexibly adjusting the screen views from full screen and gallery views. We can see up to 49 meeting participants in a single screen and totally with three screens, which is the sum of 147 participants, in the gallery view. Even though there are hundreds of individuals speaking at once, it is simple to identify who is speaking; this is because of the dynamic voice detection approach in Zoom. An additional important feature is the Live Chat, this is where all the participants in the event are able to enroll their attendance, share links, ask questions, answers the questions during the events, share their ideas with other participants.

The Zoom application's settings allow us to add our own backdrops, apply filters, watch the presenter, and ask questions without the need for additional cameras. Meetings take on a new level with more cameras [15, 16].

There are additional features in Zoom, the first is screen sharing. It makes you share your screen with other members, like host your presentations, explain your presentations, and to review work with your team members. The second feature is to make a Zoom call through which Zoom can be made as Zoom phone. Here it uses Voice over Internet Protocol (VoIP) to make a call in the cloud. Call recording and conferencing are done using Zoom phones. The additional feature is to use Zoom as a Zoom room, a virtual room with one touch sharing of information to all members in the group, at a time it is possible to use 12 whiteboards in Zoom room. The additional feature when using Zoom in a conference is to make the participants mute and disable the video, and it is possible to record the meeting [17, 18] as shown in Figures 6.8, 6.9, and 6.10.

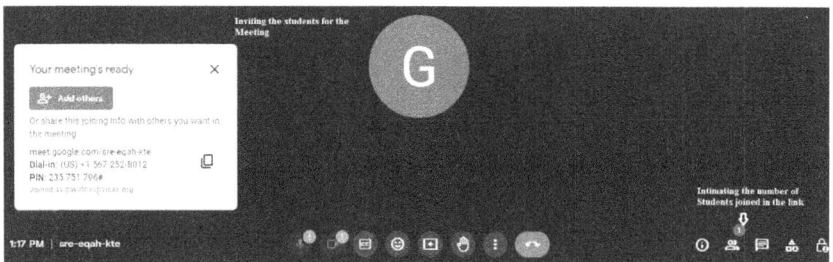

Figure 6.8   The screen showing the meeting is ready and the code for the meeting.

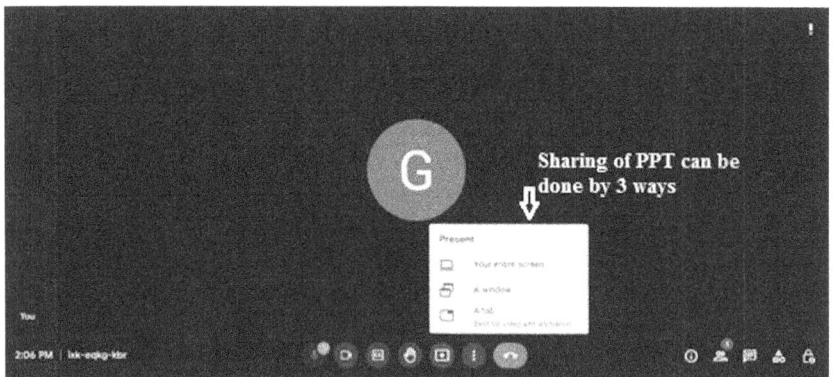

Figure 6.9   The special option, where the organiser/speaker can add and present the screen.

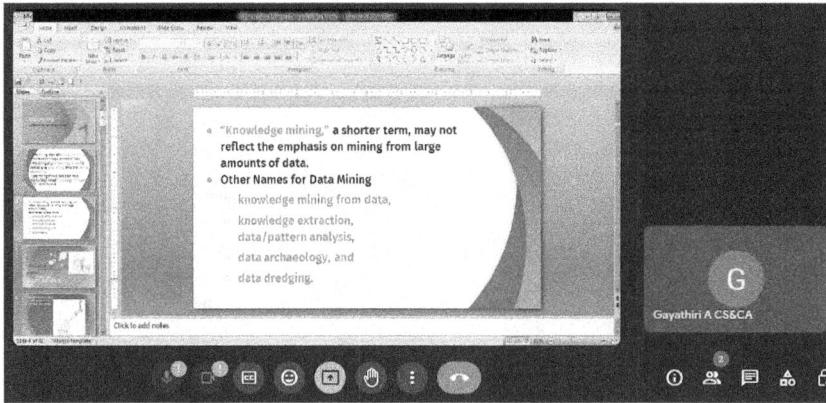

Figure 6.10   Screenshot of the speaker presenting a PowerPoint.

6.3.5 Animoto video maker tool

Animoto, a free tool for educators for creating video maker that allows for creation and sharing of videos online. On using Animoto, the key benefits of video presentations include they are easy to access at any time, study whenever you want, go at your own speed, several ways to utilise, straightforward delivery, more efficient learning, possibility for self-testing and self-study. This is the underlying principle behind why teachers utilise Animoto videos during their teaching [19] as shown in Figure 6.11.

To create Animoto video, with easy simple drag and drop the user can construct their way to fabulous videos. The initial step is to select a template, it is built into the tool. The choosing of the template is done according to our work and the students who are going to learn. The next step is to add images and videos chips into the template. The two ways by which we can add images are by uploading your own images created by you, another method of adding the images is by the Getty images preloaded in the built-in stock library. We can also specially make the video to upload in Animoto [20].

Animoto videos are produced by combination and corresponding blocks. Within any video project, the relocation of blocks is done by the trouble-free method of adding, removing and drag and drop. With the variety of styles available in Animoto, you may combine text, photographs, and video clips into a single block. It is also possible to alter the number and placement of photos and videos in your blocks by using the layout option in Animoto [21].

The option of burst block is to create a fast sequence of up to 15 photos to your videos. It is done by choosing the Burst option in the header special where you can choose several options. Here it is also possible to animate

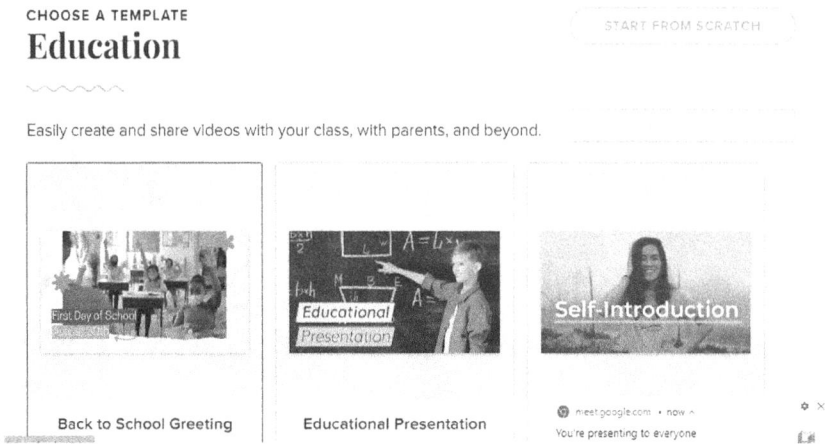

Figure 6.11   Different types of media layouts in the Animoto tool.

also. There is an icon called stopwatch, which assists you to choose to adjust the speed of burst by slower, faster, or auto.

Addition of Logo is the next essential information to incorporate your logo and get brand recognition with every view as shown in Figure 6.12 and 6.13. You can add your logo in two ways: one way is to upload your Logo and add Logo block special. It can also be added in the video either at the beginning or at the end of the video. Another method is to add a corner watermark, which will be seen throughout the video; the one thing is that it is a paid subscription. The several options for editing your blocks are Change Layout, Change Color, Add Text, Play back, and Mute, Scale, Rotate, Delete, Trim clip and Shorter & Longer is to adjust how long the video appears. The design settings are done with different Themes, Text Settings like Fonts, Colors and Sizes and Text Styles, Text color customisation, multiple text boxes, text rotation, Manual text justification and Numerical font sizing.

In video settings the options like colors, video styles, and filters are available, which makes the video look perfect. In Music there are options like mood; genre and we can select our music. Songs can be selected and made as favorite for future use. Volume control can also be done and choose from the Music Library. It is a possible and important option to add our own voice using the Voice-over option. The two customs to add voice-overs to your videos are to record in your own video in the workspace or to upload a file directly if it is already available. It is possible to add four voice-overs for a single video. Likewise it is possible to use our Animoto to make our own stunning videos.

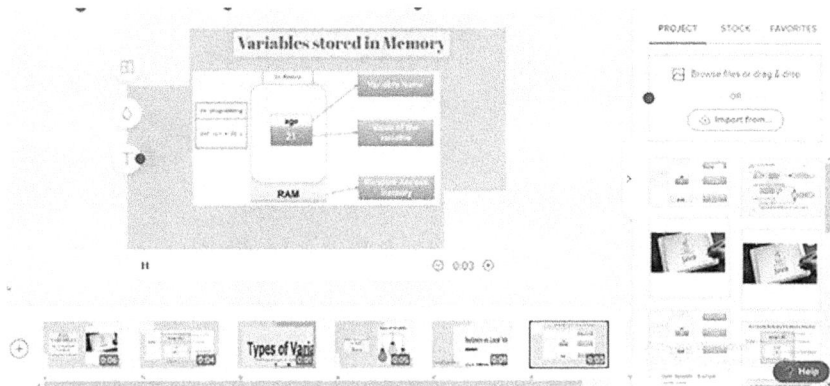

Figure 6.12 Screenshot, a video created for Java variables is playing.

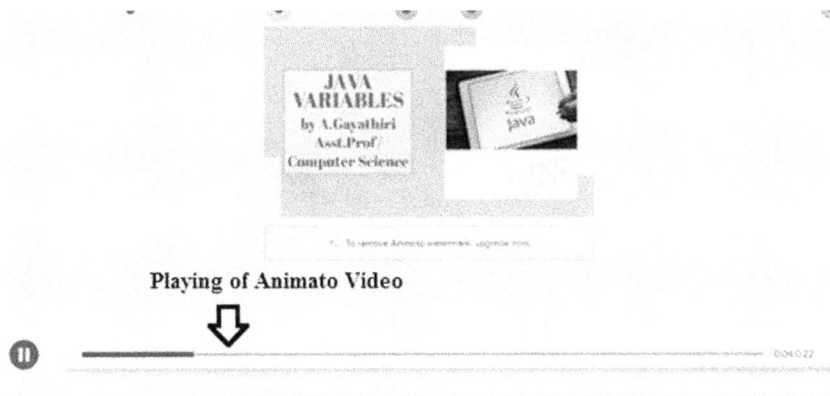

Figure 6.13 Designing an Animoto slide with photos, text, and video.

6.4 EXPERIMENTAL WORK WITH VISUAL MULTIMEDIA REPRESENTATION

The experimental work is done in the IX standard CBSE NCERT Social Science book, Chapter 2 [22], "Socialism in Europe and the Russian Revolution". This Social Science subject is not as easy to read as other subjects. It involves Year, King's name, Kingdom's name, and the different types of Revolutions in each country. The PDF of the book is available, and its visual representation of the chapter "Socialism in Europe and the Russian Revolution" is available on the Government education

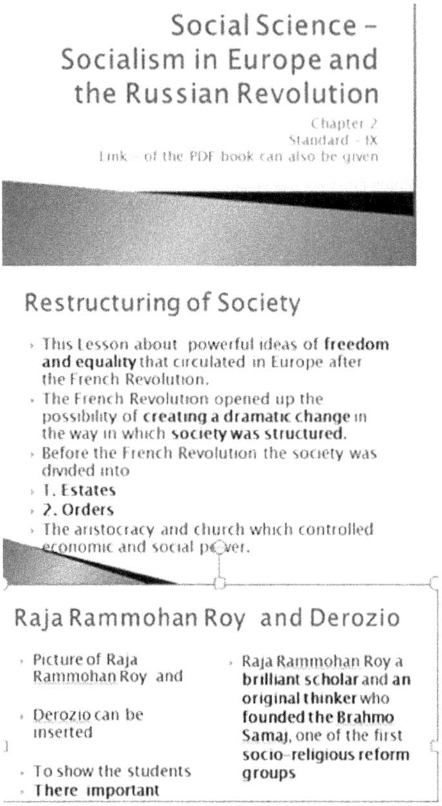

Figure 6.14   Visual Multimedia Representation (VMR).

website. Figure 6.14 shows the visual representation of the chapter by using Visual Multimedia Representation (VMR). By using the VMR the students know the concepts well. This VMR represents the images of the King, his interesting stories, and many more. This helps the students to know more much than the content. It will help the slower and medium learners to repeatedly see and hear the content, which will be memorised and remembered for a long time as shown in Figure 6.14.

6.4.1 Applications of LI-FI

A light communication system known as Li-Fi Technology is capable of sending data at fast rates spanning the visible, ultraviolet, and infrared spectrums. Several fascinating facts and helpful details concerning Li-Fi technology are as follows: Li-Fi is a form of high-speed, fully networked, bidirectional wireless data connection [23].

When an electric current is passed through the LED light bulb, a stream of light protons will be emitted from the lamp. They are capable of undergoing rapid changes in the light's intensity. By changing the luminosity races at various rates, the transmitter sends the signal, which the detector then receives. The change in light intensity is translated by this detector. Digital data is sent through communication channels, such as power ON by 1 and OFF by 0, for example.

The Li-Fi technology's important concept is that it transmits the light modulation at high speed, which is invisible by human eye. That is, the stream of data is sent continuously like 0's and 1's according to LED's ON and OFF. When the founder German physicist and professor Harald Haas demonstrated Li-Fi in a TED talk, he showed that the data is transmitted even in dimmed light rays. It uses a visible light spectrum, and infrared spectrum. It is also used underwater and its data transmission is about six thousand meters below sea level. Li-Fi has a range of approximately 10 meters [24, 25].

The various applications of Li-Fi are first is Li-Fi & Live Streaming, which is used in education fields like seminars, conferences, meetings and much more. The second is in hospitals for its safer usage. The third is Li-Fi in different wok spaces. The fourth is Li-Fi in schools for using ICT-enabled tools in teaching, the fifth is retail shops, museums, exhibitions, and in industry. Its use is continued in all areas [26].

6.4.2 Visual mode of communication

E-learning is more efficient for students to be well settled. Using computers, tablets, and smart phones it will be achieved. In urban areas the above said technical medias are well used in an efficient way as they get proper Wi-Fi signals. Wi-Fi is a signal without wires. In urban areas it is used without any interruption. In rural areas the signal is so weak towers are not so stronger to penetrate into the rural areas. To enhance the wireless signal the new technology is to be followed in that area. The technology is called Li–Fi, i.e., light fidelity as shown in Figure 6.15. In Li-Fi the signals are passed from one electronic media to the other by LED lights instead of wireless signals. Here the light rays interpret and communicate the digital signals. LED light has the capacity to transfer the digital 0 and 1 signal continuously by flickering of light. It was first developed by Harald Hass, the German Scientist, as light-fidelity transmission of data through light. It is faster than the Wi-Fi technology and by using ordinary LED lights it is possible to transmit the data. It can be used in the street lights and vehicle front and back lights for communication [27].

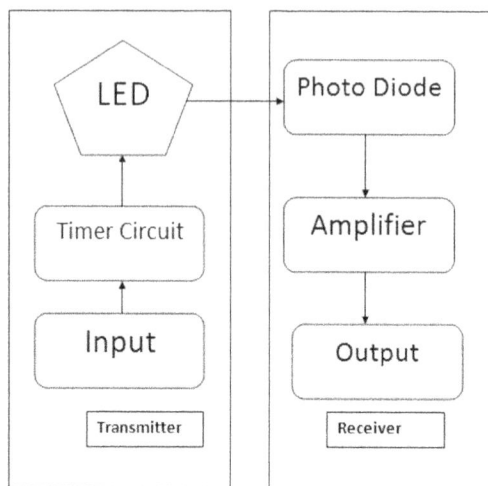

Figure 6.15   Block diagram of the Li-Fi technology.

6.5 CONCLUSION

Our goal of digital India can be achieved only by using digital and electronic equipment in the education sector for visual efficient easy interactive (VEEI) learning. This method of learning will make the students visualise the concepts that the teachers teach. It makes the students use the concepts taught while learning and during examinations also. It even gives them a hand during interview sessions for placements. To use the electronic equipment, we need the basic wired or wireless technology for transmission of signals. Wired is costly and requires manual working for implementation. Wireless is not expensive but it requires signal transmission to be enhanced in the rural areas. To enhance the wireless anywhere it can be implemented by using communication media through light as medium. The rural areas will be communicated by using Li-Fi technology. When the rural areas are equipped with digital and electronic media, the uniform education system will be achieved very fast through E-learning Visual Multimedia Representation.

REFERENCES

1. Tamil Nadu Govt Takes School Students on Education tour to Dubai by Archana R, *News18.com*, November16, 2022. www.news18.com/news/education-career/tamil-nadu-govt-takes-school-students-on-educational-tour-to-dubai-6396757.html

2. file:///C:/Users/Student/Downloads/The_Effectiveness_Of_Learning_Tools_In_Science_Lea.pdf

3. Rammurti Meena, Education System of Different Types of Schools in India, in *Journal of Advances and Scholarly Researches in Allied Education\ Multidisciplinary Academic Research*, 15, no. 9, pp. 1010–1015, 2018.

4. Students' Perception of Kahoot!'s Influence on Teaching and Learning, Sherlock A. Licorish1, Helen E. Owen2, Ben Daniel3 and Jade Li George1, file:///C:/Users/Student/Downloads/s41039-018-0078-8.pdf

5. https://ziteboard.com/whiteboard-collaboration-tool/

6. K. Varghese et al., *India and the Contemporary World-I Textbook in History for Class IX*. National Council of Educational Research and Training, New Delhi, 2023, ISBN 81-7450-536-9

7. M. Kumaresan, M.J. Basha, P. Manikandan, S. Annamalai, R. Sekaran, A. S. Kumar, "Stock Price Prediction Model Using LSTM: A Comparative Study," *2023 3rd Asian Conference on Innovation in Technology (ASIANCON)*, Ravet IN, India, 2023, pp. 1–5, doi: 10.1109/ASIANCON58793.2023.10270708.

8. www.greengeeks.in/blog/pros-cons-using-zoom/

9. mvsav.co.uk/2021-update-of-features-and-benefits-of-zoom-cloud-meetings/

10. www.indiatoday.in/education-today/featurephilia/story/india-rural-education-problems-958214-2017-01-31

11. www.trustradius.com/products/lucidspark/reviews?qs=pros-and-cons#product-details

12. 5 Ways to Upgrade the Rural Education System in India. In *India Today Web Desk*, New Delhi, July 25, 2019. UPDATED: July 25, 2019 16:52 IST.

13. Can education in Rural Sector be a Game-Changer? In *India Today Web Desk*, New Delhi, June 25, 2019. UPDATED: June 25, 2019 12:51 IST.

14. J. Muthukumari, Dr. N. Ramakrishnan, Development and Validation of E-Content on History Subject (The Beginning of Modern Age) of Ix Standard Students" in the *International Journal of Research-Granthaalayah*, 5, no. 9 SE, 2394–3629, September 2017.

15. M.R.M. VeeraManickam1, M. Mohanapriya2 and Debnath Bhattacharyya International Journal of u- and e- Service. *Science and Technology*, 9, no. 8, pp. 111–132, 2016.

16. R.K. Dhanaraj, S.K. Islam, V. Rajasekar, A Cryptographic Paradigm to Detect and Mitigate Blackhole Attack in VANET Environments. *Wireless Networks*, 28, no. 7, pp. 3127–3142. 2022.

17. Dr. Lutz Laschewski "Innovative E-learning in Rural Areas: A Review" Article in *SSRN Electronic Journal*, March 2011.

18. Article by Ritesh Rawal, Founder, Dudes and Dolls, The Cosmic School, Faridabad, What is the rural education scenario in India and how can we change it?, *India Today Web Desk*, New Delhi, August 5, 2019. UPDATED: August 5, 2019 15:31 IST.

19. Dr. Nachimuthu Periyar University, " Need of e-content development in Education ", *Education Today, An International Journal of Education & Humanities*, Vol. 3, no. 2, pp. 72–80. APH pub, New Delhi, July-Dec 2012. ISSN: 2229-5755

20. R. Shankar, B.K. Sarojini, H. Mehraj, A.S. Kumar, R. Neware, A. Singh Bist, Impact of the Learning Rate and Batch Size on NOMA System using LSTM-Based Deep Neural Network. *The Journal of Defense Modeling and Simulation*, 20, no. 2, pp. 259–268, 2023. doi:10.1177/15485129211049782

21. Suresh Kumar Arumugam, Amin Salih Mohammed, Kalpana Nagarajan, Kanagachidambaresan Ramasubramanian, S. B. Goyal, et al., "A Novel Energy Efficient Threshold Based Algorithm for Wireless Body Sensor Network", *Energies* 15, no. 16, pp. 1–19, 2022.

22. Dr. Gopikumar Shivaramaiah," Teaching Learning Methods: Traditional vs. Modern vs. *Peer-Assisted Learning*, June 01. 2018.

23. Shu-Sheng Liaw, "Investigating Students' Perceived Satisfaction, Behavioral Intention, and Effectiveness of E-Learning: A Case Study of the Blackboard System", *Science Direct Journal*, Received 25 July 2007; received in revised form 29 August 2007; accepted 4 September 2007.

24. https://animoto.com/blog/guides/getting-started-animoto

25. TNN/Updated: Feb 18, 2021. "Two Gujarat Villages First in India to Get LiFi: Ahmedabad News–Times of India." *The Times of India*, TOI, Timesofindia.indiatimes.com/city/ahmedabad/2-gujarat-villages-first-in-india-to-get-lifi/atcleshow/81080015.cms

26. A. Suresh Kumar, S. Jerald Nirmal Kumar, Subhash Chandra Gupta, Anurag Shrivastava, Keshav Kumar, Rituraj Jain, et al., "IoT Communication for Grid-Tie Matrix Converter with Power Factor Control Using the Adaptive Fuzzy Sliding (AFS) Method", *Scientific Programming*, 03, pp. 1–11, 2022.

27. M. Mallick, M. K. A, S. Govindaraju, A. S. Kumar, M. Kandasamy, P. Anitha, "Analysis of Panoramic Images using Deep Learning for Dental Disease Identification," *2023 Third International Conference on Artificial Intelligence and Smart Energy (ICAIS)*, Coimbatore, India, 2023, pp. 1513–1517. doi: 10.1109/ICAIS56108.2023.10073939

Chapter 7

Towards the 21st century
Solutions to the problems and pitfalls of mobile learning in education and training

P. Sumitra[1], S. Sabitha[1], M. Sathiya[1], G. Sathya[1], A. Gayathiri[1], and M. Kumaresan[2]
[1]PG and Research Department of Computer Science and Applications, Vivekanandha College of Arts and Sciences for Women (Autonomous), Namakkal, Tamilnadu, India
[2]School of Computer Science and Engineering, Jain (Deemed to be University), Bengaluru, Karnataka, India

7.1 INTRODUCTION

Teaching strategies and learning models have evolved as a result of the preface of information technologies and the Internet in learning. Learning patterns are constantly knocking on students' or trainees' doors. Anyone may learn today, anytime, anywhere in hand with updated information at affordable cost. Today's world is becoming more mobile. Mobile and communication technology make a major impact on people's lives and fundamentally change our tradition and society. Cellular gadgets give access to enormous information and allow messages virtually anywhere. Learning approaches have also shifted in response to these technologies. Mobile devices are an essential part of our day-to-day lives. Demand for mobile phones surged as their prices fell. Mobile technology use is greater than ever before in industries, including banks, business, tourism, amusement, information centres, and more [1, 2]. People have the opportunity to grab the soft or hard skills by using mobiles at their own pace. Mobile devices are used by students to receive teaching in a traditional classroom. The sensors and other features are integrated into learning objects for the learning environment. They identify the learners' current context and other learning objects. The learner is served on the basis of context and situation. Technology has delivered information between students, tutors, and instructors. Education now takes place outside of the traditional classroom as well. Although traditional education is still practised, the phenomenal growth and expansion of ICT opened the door for the quick rise of Internet-based education. Our higher education sector has been drastically altered by the Internet. Millions of students now have the opportunity to acquire their education through Internet-based instruction.

DOI: 10.1201/9781003362685-7

7.1.1 Objectives

Studying the pioneering methods of M-learning in education. We find a remedy for the problems and pitfalls with respect to M-learning in teaching and training.

7.1.2 Scope of mobile learning

The mobile learning is a way where a learner can access education at any location and time by using handy learning equipment. Any form of education that makes use of the learning opportunity offered by technology of mobile is referred to "M-learning." M-learning is practical in that it offers access to all of the many learning resources available and can be accessed from almost anywhere. Additionally, it is collaborative; everyone utilising the same content can share nearly instantly, allowing for the quick exchange of criticism and pointers. M-learning improves portability further by substituting portable devices with customised learning content for books and notes. Mobile learning offers a number of fantastic benefits, but when M-learning is poorly structured, issues can arise.

Designing efficient mobile learning will get simpler as mobile devices gain in power. With mobile learning, you can learn wherever you are without having to be physically connected to a network. With the most recent e-learning technology, there is also a focus on "just in time," "just enough," and "just for me" notions [3]. The M-learning system can offer location-dependent educators using GPS technology. There are several devices used by students in the M-learning process. They are as follows.

7.1.3 Types of M-learning

Note book computers – They offer desktop computer-like features, on the one hand, but are also portable and support wireless connectivity, on the other. They continue to have high prices.

Tablet PC – They offer desktop computer-like features, on the one hand, but are also portable and support wireless connectivity, on the other. They continue to have high prices.

Personal digital assistance (PDA) – They are compact and have powerful processors. The two most popular operating systems are Pocket PC (Microsoft) and Palm.

Cellular phones – Primary functions of old mobile devices are voice communication and text message sending and receiving (SMS). Low data transfer rates and limited memory are a few of their drawbacks. GPRS or WAP technology can be used to connect higher-class cell phones to the internet. Their cost keeps dropping with time.

Smart phones – These hybrid gadgets have the feature combination of PDA and mobile phone. They are larger than cell phones and smaller than PDAs. Windows Mobile, Symbian, or another operating system can run because they have web browsers. They could be effectively applied to mobile multimedia learning. Modern mobile devices use a variety of communication mechanisms. Their skill ranges widely.

Mobile communication as a global system – This is the best digital cellular systems. It is most commonly utilized mobile network in the world, being used in more than 100 nations. It offers integrated fax, paging, high-speed data, and voice mail. Of all the digital wireless standards now in use, it gives the finest voice quality.

Wireless application protocol (WAP) – This wireless communication protocol is open-source and free. It enables the development of sophisticated communication services and cellular phone access to websites.

General packet radio service – By applying a technique it is possible to communicate data at fast speeds through wireless networks and other media GPRS offers approximately four times the speed of a traditional GSM system.

Bluetooth – Short-range radio technology is wireless technology. With Bluetooth, signals may be sent over short distances between phones, laptops, and other devices, making communication and synchronisation easier.

IEE802.11 – Wireless local area networks use this particular radiotechnology (WLANs).

The chapter is organised as this, Section 7.2 gives an overview of research on mobile learning, Section 7.3 aims to discuss technological advancements in education, Section 7.4 describes the fundamental components of mobile learning, Section 5 explains the fundamental features of M-learning, Section 7.6 explains the advantage and disadvantage of mobile learning, Section 7.7 discusses the internal and external challenges to technology in education, Section 7.8 discusses the data collection tool and finally Section 7.9 concludes the chapter.

7.2 MOBILE LEARNING RESEARCH OVERVIEW

An article review is an evaluation of studies that has been found in the literature and is relevant in a researcher's area of study. The literature should be described, concluded, and assessed to make a clear evaluation. The researcher helps to determine the theoretical framework of their own investigation. There were several articles relating to mobile learning found. These articles cover a range of years, from 2008 to 2020, which is in Table 7.1.

Table 7.1 Researcher overview in M-learning

Reference number	Year	Author name	Objectives of the study	Findings of the study
[4]	2008	D. Mcconatha and M. Praul	The purpose is to assess the potential performance, challenge and possible outcomes in an organisation. Additionally, it offers the effectiveness of mobile learning in a lecture room environment.	1. There is a significant difference between the experiment and control groups. 2. The standard deviation (SD) results were nearly two times greater (11.7 vs. 6.1) than that of the control group. On average 89% better than those who only used printed. 4. The results so as to the students in the class who used cell phone to study exam materials and perform better than those who used more traditional methods.
[5]	2009	S. Paliwal, & K.K. Sharma	To investigate the issue with mobile learning's integration into traditional education and training. Find out how M-learning can be incorporated into traditional education. To develop adaptable teaching solutions that will promote learning in a range of contexts and provide access to information via various devices.	1. Its conclusion is the telecommunications operators do not consider mobile learning to be a reliable source of income. 2. This presentation has addressed the issue of mobile learning's Integration into traditional education and training.
[6]	2010	M. Osman, El-Hussein, & C.J. Cronje	To explain the significance of mobile learning by using examples from post-secondary education to illustrate its fundamental ideas. To disassemble the model's core components and comprehend in context of advanced education.	1. All erudition that takes place in different locations for learner, learner mobility, and learning mobility are included in the definition of mobile learning. 2. Despite the fact that neither researchers nor designers themselves are totally certain of the enduring effects of this medium, applications and uses in mobile learning have grown in a variety of settings. 3. Mobile learning is more effective when the technologies are properly designed. 4. It's also important to go into detail about the many benefits and drawbacks of using mobile learning tools to deliver higher education.

[7]	2011	O.N Keskin and D. Metcalf	To comprehend and discuss modern theories and views in M-learning.	1. Similar to researches, the field of mobile learning has a bright future. 2. Researchers have discovered that there are numerous diverse mobile Learning methodologies, ideas, and practices.
[8]	2012	Z. Taleba and A. Sohrabi	To assess how gender and educational level affect the rate of mobile use in education to research the impact of educational organisation on how frequently students utilise mobile devices for learning. To determine the elements most responsible for the rate of mobile learning uptake.	1. It has also been revealed that many students utilise their mobile phones for educational purposes. 2. The findings conclude the percentage of university students using their mobile phones for educational purposes. 3. Additionally, there is no difference between academic courses and students usage of mobile phone for educational purposes. 4. The findings of the present study demonstrate that cell phone usage is pervasive and consistent across all academic disciplines.
[9]	2013	F.S. Narayanasamy and B.K.K Mohamed	To ascertain the students' level of familiarity with mobile learning and the elements. Use of mobile learning in the classroom and the disclosure of mobile learning services were easily scrutinised.	1. The outcome showed that students are prepared to accept mobile technologies learning process and those they have sufficient knowledge and awareness about mobile learning. 2. This outcome also provided an indicator of what students expected from higher education institutions' mobile phone-based university services. 3. The research revealed that university possessed infrastructure and use of m-learning services. 4. Results showed that the students have the knowledge and awareness necessary to use these tools in their learning. 5. The report states the best features and benefits of using mobile technology for education services was giving learners quick access to information, regardless of where they are located.

(Continued)

Table 7.1 (Cont.)

Reference number	Year	Author name	Objectives of the study	Findings of the study
[10]	2014	T. Bansal and D. Joshi	To investigate how students perceive WhatsApp's omnipresence in mobile learning. Researching the educational advantages of WhatsApp mobile learning. To examine the social interactivity in the mobile learning environment of WhatsApp. To investigate students' attitudes about mobile learning with WhatsApp.	1. Studies demonstrate that many stakeholders in education have a favourable attitude toward the application of mobile learning in classrooms. 2. Only a small portion of students who are married feel that learning at anytime and anywhere interferes with their family life. 3. Understudies feel way more comfortable in confronting their quick criticism to the issue, having more chance to memorise and have clarity on issues, talking about past learned subjects and at long last have a steady get to in the learning fabric. 4. Generally, learner states of mind toward WhatsApp m-learning are positive. 5. They discovered WhatsApp m-learning to be an effective teaching strategy, a helpful learning tool, and a highly individualised educational medium. 6. Students prefer WhatsApp mobile learning to traditional classrooms and use social media for teaching and learning in the future. 7. This think about assist bolsters this, as understudies report that considering through WhatsApp m-learning makes their life simpler since they can learn at any time and from any area.
[11]	2015	A. Saxena and A. Saxena	To compare student perceptions in the future, India's education sector according to gender. To gauge how students generally feel about the prospective for mobile learning in India's education sector.	1. The results conclude that there were no statistical significant gender differences in the students' opinions about the future of mobile learning in India's education sector. 2. The vast majority of students exhibit a significant amount of favourable opinions toward M-learning.

| [12] | 2016 | M. Al-Emran, M.H. Elsherif and K. Shaalan | To investigate the approach of students and teachers on the use of mobile learning in higher education institutions in Oman and the UAE. | 3. The discoveries appear that there's broad understudy understanding with the advantageous angles of M-learning and its positive thought as a viable learning apparatus.

1. Score of the average students' attitudes overall was determined to be (3.43), and their average score for how valuable they alleged using cell phone devices to be for their studies was (3.27).

2. It was found that there was acritical contrast within the states of mind of the understudies toward their possession of smartphones, with the contrasts favouring both gadgets (smartphones and tablets).

3. The findings showed a statistically significant variation in the students' attitudes depending on where they lived, with the disparities favouring students from the United Arab Emirates.

4. The findings showed that the pupils' attitudes toward their age varied statistically significantly.

5. The findings showed that 99% of pupils had cell phone devices such as smartphones or tablets, and only 1% do not.

6. It was discovered that while 16.7% of students used cell phone devices for learning purposes, 41.5% of them used them for email and web browsing on their mobile devices (smartphone or tablet).

(Continued) |

Table 7.1 (Cont.)

Reference number	Year	Author name	Objectives of the study	Findings of the study
				7. The findings showed that M-learning is well received by both students and teachers of both sexes. 8. Findings showed that practically all educators have favourable sentiments toward M-learning, regardless of their academic standing, academic experience, or ownership of smartphones.
[13]	2017	K. Fouzdar and K.S. Behera	The author make a consideration in Sidho-Kanho-Birsha College in India's, West Bengal basically PG under studies feel around flexible learning. He examines how both the sexes feel about flexible learning and compares the acknowledgments of versatile learning among the PG understudies. To assess how SC/ST and PG General students in the Purulia District feel about mobile learning. To assess how students in the Purulia District's PG Arts and Science streams feel about mobile learning. To assess how Purulia District PG students in the second and fourth semesters feel about mobile learning.	1. According to the report, the Post Graduate has a good or average toward mobile learning, neither more favourable nor unfavourable for mobile income. 2. The result was found that there was no perceivable distinction within the demeanors of PG male and female understudies toward versatile learning at Sidho-Kanho Birsha College. 3. Findings show that they have similar attitudes about mobile learning in both urban and rural settings. 4. The study's findings revealed no appreciable distinction between PG SC/ST attitudes of the students and those of PG general students. 5. According to the survey, many of students have different attitudes regarding mobile learning. 6. Additionally, it was discovered that there were no appreciable differences in the perspectives on mobile learning.

7.3 ADVANCES IN TECHNOLOGY IN LEARNING

New technologies are being incorporated into education and knowledge techniques. To humanise the process of teaching and learning, several different resources are used. In order to generate new behaviours, skills, attitudes, preferences, or understandings, learning can include mixing different types of knowledge. When being taught with ICT tools like a blackboard, an overhead projector, a video projector, and Internet access, students participate and learn from one another.

However, the main shortcomings of the conventional method are the need of efficient classroom equipment, accessibility of the location, and the small number of classrooms. Due to the fact that students may access the learning resources on these entire internet learning environments at any time and from any location, mobile learning has become more and more popular at many educational institutions in recent years.

Mobile learning can get instruction from any area at any time by utilising convenient learning hardware. Any type of education that makes use of the opportunities for learning provided by mobile technologies is referred to as "M-learning," sometimes known as "mobile learning." M-learning allows you nearly unlimited access to all of the available learning resources. By using portable devices in place of books and notebooks, M-learning is practical, highly portable, and increasingly capable to create efficient mobile learning.

7.3.1 Reason behind the development of mobile learning

M-learning refers to the capacity to move through course material on a person's own personal device. The potential of anytime, anywhere is introduced by mobile learning, which offers its users a wealth of advantages. M-learning has been embraced by numerous organisations, enhancing user engagement and producing knowledgeable workers. The reason behind the technological development of M-learning is as follows [14].

- It facilitates flexible and accessible learning.

  Learners have the greatest accessibility and freedom with mobile learning. Cellular phone devices allow learners to access M-learning courses any time and any location. They can access and take the courses from home, on the bus, or even in their free time. One of the problems with e-learning was that it had to be completed during business hours, when staff members are already quite busy due to a variety of scheduled activities. Implementing M-learning is the ideal solution to solve that issue.

- It is just-in-time learningOur work life is becoming more hurried as a result of technology. Every challenge, procedure, and timed learning technique has been improved. Today, technique can offer staff just-in-time training, making it ideal for this fast-paced world. Employees can easily access learning modules from brief mobile learning courses, or "micro learning," to review material or pick up new skills while at work. For instance, how to act in a meeting or how to respond to a consumer the right away.

- It increases completion rates

    No longer does anyone have the time to sit through 30-minute e-learning courses. Because M-learning courses are short and to the point, students can finish them fast and go onto the next. Due to the reduced daily time commitment, this raises the completion rates of these courses. Employees can be more motivated to advance through and complete their training by adding gamification components to micro learning courses, such as prizes, points, and level progression.

- It offers higher engagement

    The modern person is accustomed to getting information on their smart phones in brief, targeted bursts.

- It offers personalised learning

    Since L&D professionals realised they needed to adopt and carry out a plan for a learner-centric approach to learning by providing learners with freedom and accessibility, M-learning has grown in popularity. Another learner-centric strategy is personalised learning, which works well in M-learning courses and provides instruction tailored to each learner's specific needs, interests, and strengths. This is accomplished by using learners' own learning paths, which respond to or adjust based on their development, motivations, and goals through the gathering of precise and pertinent data.

- Educational technology and mobile learning

    Mobile learning and educational technology are closely related. Today, educational technology is essential to students' success since it provides them with all the resources they require for the optimal learning experience. It is possible for us to learn whenever we want, wherever.

7.4 BASIC ELEMENTS OF MOBILE LEARNING

Learner, teacher, environment, material, and assessment are the fundamental components of mobile learning. Basic components of a successful mobile learning technique are shown in Figure 7.1. They need to be well-organised

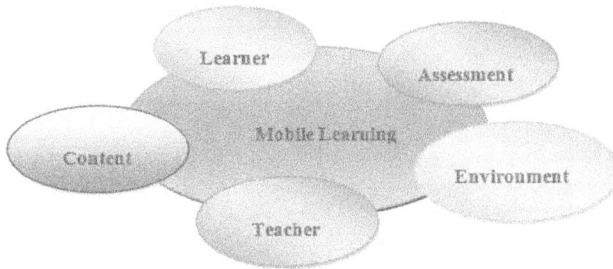

Figure 7.1 Fundamentals of mobile learning.

for mobile learning to be successful in order to increase efficiency and success. In order for the student to get interested in what is being taught, they should also be engaging and fun to use. M-learning has become crucial for higher education institutions all over the world due to its capacity to deliver educational information at anytime, anywhere, and with greater student engagement.

7.4.1 Learner

The first priority is always the student or learner. Whatever the method of instruction, its objectives can be achieved without the development of students. From setting the goals until the evaluation step, the student actively participates. A learner's role is as follows:

- Obtain information as needed.
- Being liable for one's own education.
- Adapting to their rate of learning.
- Find and apply their preferred learning methods.
- Create and distribute fresh content or products.
- Together with their classmates, they study.
- Assessing both their own group and other groupings.

7.4.2 Teacher

In traditional learning contexts, teachers transmit information to pupils while storing it in books and other media. On the other side, modern information storage methods support more readily accessible information for pupils. Modern information technology and the role of teachers is shown in Figure 7.2.

Roles of teachers in mobile learning: capable of using necessary mobile tools and technology.

- Analyse the advantages and disadvantages of the already employed techniques and look at ways to address the shortcomings.

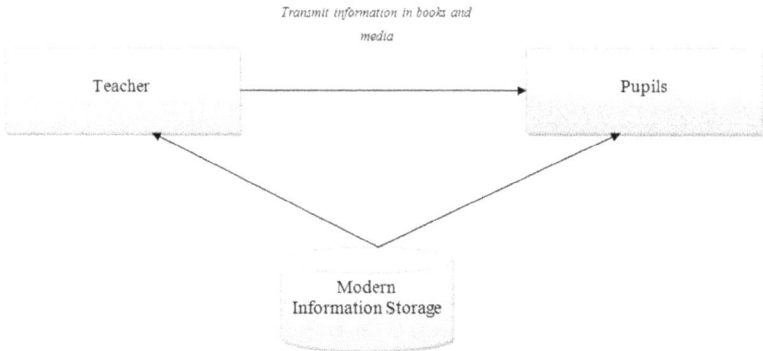

Figure 7.2   Role of teacher in modern information technologies.

- Facilitator's manual.
- Counseling.
- High degrees of confidence regarding courses.
- Gain knowledge with your students.
- Removes obstacles.
- Improve student enthusiasm.
- Plan activities that will encourage group collaboration and engaging interactions.
- Plan actions for process evaluations.

Before television, teachers' primary responsibilities were to act as subject-matter experts and disseminate information to their students. The transformation in media formats transformed the typical teacher's position from one of expert to one of presenter of the knowledge of others.

7.4.3 Content

All interested parties, including students, instructors, parents, and others, should be consulted when choosing the content. Otherwise, educators won't receive the required outcomes. A user must be able to swiftly focus on the information they need when using learning content. Additionally, interactive games and tests can be used to communicate the content. Content should include multimedia elements like graphics, video, and other.

7.4.4 Environment

Student involvement and academic success are directly impacted by the educational environment. An opportunity to construct the environment by fostering learning to educators via a mobile learning method. With the availability of mobile technologies, students taking classes in person can access both the material covered in class and supplementary material online. Students can access course material whether in a coffee shop or on a train. The environment needs to promote more contact between students and between students and professors. Blogs, social networks, and wikis can all be used to foster more social engagement. Cell phone devices such as laptops, smartphones, and other mobile tools must be accommodated in these contexts.

7.4.5 Assessment

The full M-learning process must include assessment. Mobile devices can evaluate, record, and inform instructors of student performance. Therefore, student evaluation should be done using software, chat rooms, discussion boards, databases, online tests, or evaluation of project. Learners should assess both them and other people. It offers components required precisely to assess a learner's knowledge, talents, creativity, etc.

7.5 BASIC CHARACTERISTICS OF MOBILE LEARNING

Mobile learning has unique qualities. Mobile learning's primary characteristics are its pervasiveness, the portability of its instruments, its blend of private, interactive, collaborative, and instantaneous information. Figure 7.3 shows the distinctiveness of M-learning.

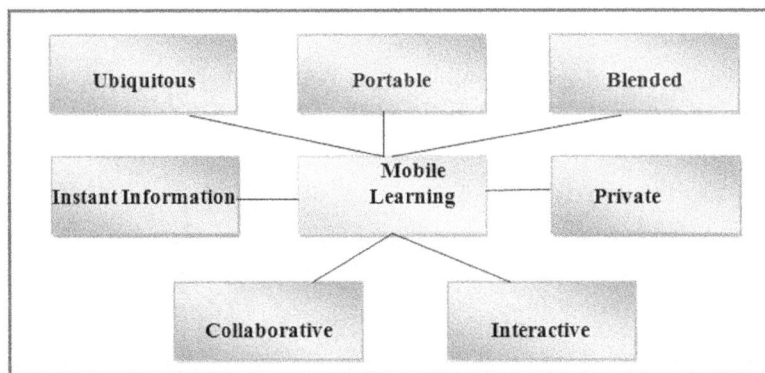

Figure 7.3 Characteristics of M-learning.

Compared to other learning styles, M-learning is more spontaneous [15]. The greatest distinguishing quality of mobile learning is likely its spontaneity. Because M-learning is context-sensitive, learners can learn anywhere. Teaching and learning are being revolutionised by wireless technology, which transforms traditional classroom-based education into education that can take place anywhere and at any time. Examples include laptop computers, palmtop, personal computers, and cellular phones.

7.5.1 Portable

Tools for mobile learning are compact and transportable. It is available to students at all times during their educational activities [16].

7.5.2 Blended

The advantages of many learning domains, including M-learning, E-learning, face-to-face learning, and related learning, are combined in blended learning, an omnipresent learning solution [17]. Virtual and physical learning materials are effectively merged in blended learning to offer a variety of learning modalities.

7.5.3 Private

Mobile learning is confidential. Typically only one learner at a time has the provision to access the mobile tool, and when students need to access material, they connect and download on their own without the assistance of other learners [18].

7.5.4 Interactive

The learners can effectively communicate with teachers, peers, and experts successfully using a variety of media [19].

7.5.5 Collaborative

Mobile devices let professors and students communicate with one another. Therefore, collaborative learning activities using mobile technologies may be used in education [20].

7.5.6 Instant information

Everything about using a mobile gadget is immediate. It asserts that there is a demand for immediate responses to particular queries. This criterion must be reflected in learning materials by including information that makes it easy for students to focus on the lesson at hand. Definitions, formulas, equations, and other types of quick knowledge are examples [21].

7.6 MERITS AND DEMERITS OF M-LEARNING

Since the early 2000s, mobile learning has existed in some form or another. This was a natural progression given how pervasive technology is in our lives nowadays. Accordingly to personal computer revolution, educators and learners are increasingly using electronic devices for learning and teaching. As time passed, mobile devices that were more portable became popular. This results in students studying presentations on their tablet, watching an instructional video, taking a multiple choice quiz on their mobile phones, or even joining massive online open courses with thousands of other students all over the world to keep learning. Below shows the advantages and disadvantages of mobile learning.

7.6.1 Merits of M-learning

- Anytime and anywhere learning

  Students are not required to learn in a classroom. Students can visit classrooms at any time to revise coursework. No matter where you are or what time of day it is, learning is now available to you. Instead, students can pursue independent learning while on the go. By doing this, kids participate voluntarily and on their own terms, which makes the learning process feel more powerful to them than just another day at school.

- Abundance of knowledge

  It has to do with how much information is available online. All of this data can fit into a strawberry-sized piece of paper. However, the number of pages in books and other written sources of knowledge is restricted. Because of the amazing technological advancement known as digitisation, you no longer need to carry about heavy books and notes. They can easily find what they need on their smartphones and laptops.

- Portability

  Information follows you wherever you go, and since the internet currently contains an infinite amount of information, you may carry all the information you require with you on a single device.

- Knowledge is free for everyone

  Every user may easily understand the concepts thanks to the content's availability on the web in the audio, video, written, and graphical formats. Additionally, since 2000, as mobile technology has advanced, so too have other technologies, such as software, cameras, recorders, and other electronic devices. Together, they have put a lot of effort into giving users accurate information in the most user-friendly formats. To access knowledge from the internet without spending a dollar, all you need is a device and an internet connection.

- Define your own pace

  You can conduct independent R&D to broaden your knowledge if you have already selected your interest area and have access to many information sources. You are always free to learn at your own pace. Anyone can consume as much information as they want and still benefit from learning. The amount of information is limitless. The key to maximising your learning advantage is to select your field and begin acquiring as much information as you can at your own speed. Just keep in mind that any innovation requires extensive study and research before it can be put into practice.

- Mobile learning significantly helps in education

  Many parents have moved to mobile learning during a shutdown as most institutions have shifted to e-learning and work from home during the epidemic. Children can learn more effectively without teachers by using the tools at their disposal and taking notes. In order to avoid having their curriculum disrupted, mobile learning has enabled schools and institutions to connect with students at their homes.

7.6.2 Demerits of mobile learning

Even if mobile learning has advantages, there are drawbacks to its application and the techniques used. Although it gives you a wealth of knowledge, there are several restrictions to mobile learning that must be taken into account. Information accuracy is the main problem with mobile learning.

- Extended screen time is the major drawbacks of mobile learning.

  Increased screen use increases health hazards such as eye strain, behavioural disorders, and sleep problems. Mobile devices and computers have replaced time spent reading books, thus spending time online has drawbacks of its own.

- Lack of physical interaction

  Online learning and discussion are both possible with mobile learning. However, it lacks the sensation of in-person conversation and physical connection, which both require bodily spontaneity. Children tend to pay less attention in online classrooms since the natural environment established in a school classroom can never be recreated. Children may also begin performing poorly as a result of these negative aspects of learning.

- Efficiency

  The amount of knowledge you read and comprehend is the most crucial factor in learning. Despite the fact that there is an infinite amount of knowledge available, learning and understanding a subject effectively requires a certain setting and configuration. Most essential, your ability to learn from a book or the internet is driven by your desire.

- Major limitations of M-learning are too many distractions

  Every time a notice appears on your screen, it distracts you and interferes with your ability to study. For instance, it's challenging for lecturers to attend to pupils who are causing mayhem in the classroom. Similar to how when you close adverts and notifications, information in any form, including written and video content, is temporarily halted. Even email and social media can cause constant distractions while students are trying to concentrate.

- Dependence on technology

  Users are more reliant on technology than ever, even if M-learning and online education are objects of the present period. The use of mobile devices has created health issues in addition to security risks. There are less chances of missing out on your favourite songs, social media, or communications if you put your phone aside for the day. Why not take up gardening or go for a walk outside?

7.7 EXTERNAL AND INTERNAL CHALLENGES TO TECHNOLOGY IN EDUCATION

7.7.1 Technology challenges in education

The first-order obstacle to effective classroom technology integration is the external factors that are outside the instructor's control. External hurdles must be addressed by institutional changes, and most advancements take time.

- Access

Early studies of technology paid a lot of attention to expanding the number of computers in classrooms. The efficiency in computer labs is not feasible if it is restricted to one hour per week. Even though one-to-one computing class is implemented many lack frequent and dependable access to computers. Due to inconsistent access to computers, it makes it quite challenging for teachers to incorporate the lessons. Exposing students to regular classes with software, hardware plays a vital role [22].

- Training

Lack of training and professional development is the most common reason for not adopting technology in the classroom. Given that technology is ever-changing, teachers today report feeling more comfortable using software, the Internet, and classroom technology. More than ever, educators must stay current with their technology knowledge [23]. Countless new technologies will be produced during teachers' careers, necessitating additional training to maintain their skills up to date, even if a school district hired solely instructors who were knowledgeable in the most recent classroom technology. Schools and locale will proceed to point to inadequately proficient advancement as a critical deterrent to innovation selection within the nonattendance of the subsidising required to supply continuous innovative preparing. School administrators should look for support to identify and give continuing training in order to attain effective technology integration. Student curriculums that incorporate technology, professional development tools, and evaluations are all approved by the International Society for Technology in Education (ISTE) as meeting their standard technology integration in the classroom. Programmes for teacher technology skill development that have been approved by the ISTE include in-person training, online courses, communities of learners online, online learning modules, and in-class mentoring. In certain school districts, professional learning communities are led by master teachers who have had experience using educational technologies. These groups meet frequently to receive training and support for technology integration. At long last, instructive innovation analysts and computer programme engineers ought to be looked for by schools and teaches for preparing. Various computer programme firms give teaches with free online preparing, proficient advancement courses, and continuous back.

- Support

The process of implementing new instructional technologies can take some time. If a technology is implemented across the entire school, teachers should have access to ongoing support from qualified experts rather than a single hour-long meeting before the start of the school day. The process of implementing new instructional technologies can take some time. If a technology is implemented across the entire school, teachers should have access to ongoing support from qualified experts rather than a single hour-long meeting before the start of the school day. Teachers will have access to the resources they need with high-quality support from both educational technology developers and school personnel. The understanding that support is easily accessible may improve support for classroom technology.

7.7.2 Internal challenges in education

The educators are highly responsible in use of technology. This section will include issues relating to teachers, their perspectives, and their ongoing professional development. Teachers are committed to the traditional approach. The incorporation of technology may be hampered by instructors' opposition, which is intimately related to attitudes and beliefs. Finally, we discuss guidelines, technical knowledge, and expertise.

- Teachers attitude and beliefs

  Teachers have a big influence on students' attitudes and opinions about using technology in the classroom.
  The educators use the technology to discuss pedagogy in the classroom [24].

- Confidence in skills and knowledge

  Given the wealth of instructional technology accessible, it is crucial that teachers are at ease with and confident in their abilities to use it. While kids nowadays are grown in a world dominated by computer technology, many current teachers grew up without access to tools like the personal computer or the internet. Teachers may feel less in control of the classroom, use less technology, and be less likely to explore novel uses for technology if they feel they lack the requisite abilities [25]. Teachers who are less computer literate can preserve control in the classroom and hopefully avoid getting prepared to face the challenges of teaching digital natives in a digital environment by sticking to traditional teaching methods.

7.7.3 Technology and learning

Curriculum can integrate technology. Teachers have access to a variety of technological tools. The activities in the classroom will be governed by the chalk and talk teaching strategy. To effectively employ computers in the classroom, teachers must make the transition from the teacher-centred classroom to the student-centred classroom.

Trainers must follow shift basis from teacher-centred to student-centred to effectively use the computers in the classroom [26]. Numerous classroom activities will be carried out by students using the chalk and talk teaching approach. Therefore, it is likely that technology will only be used in supplemental demonstrative activities within specific educational modules [27]. With the aid of educational technologies, the teacher can act as a facilitator for the student's participation in more learning activities.

- Teacher resistance to technology in the classroom [28]

    It is evident from perusing online teacher forums that integrating new technologies into lesson plans can be challenging. The fact that many teachers are happy with their current lesson plans is probably the most frequent excuse given by educators for why they are not aggressively integrating new technologies. Classroom instruction is driven by a teacher's desire for their students to learn efficiently, and if present lesson plans suit those demands, there isn't much incentive for the teacher to change them. There are undoubtedly many reasons why a teacher might be reluctant to introduce new technology into the classroom, but if they make the decision to do so, teachers must first pick which technologies to use.

    It can be difficult for teachers to select the Internet technologies, tutoring programmes, and learning settings that will improve student learning and fit with curricula because there are so many options available. Many technologies make claims that they would enhance students' academic and cognitive abilities. However these claims might be incorrect and are frequently made solely for commercial purposes. Perhaps as a result, choices regarding technology are sometimes taken by district or school administrators without consulting teachers. This can be advantageous in some ways, such as saving teachers the time and energy needed to evaluate technologies, but a lack of options can also have a negative effect on how an instructor views the technology. Although the technology may actually make teaching easier and more fun, teachers may perceive it as an imposition.

- Increase acceptance of classroom technology [29]

    We suggest a number of strategies that teachers and academics can use to encourage technology integration right away because it is certain that more and more students will accept classroom technology over time. The crucial things for the teachers are in choosing the technologies for handling their lessons. The most important decisions for teachers to make when handling their classes with technology are these. Since every teacher's experience of teaching is unique and different, they thought that if they used technology, they would lose their ability to educate in the way that suits them best. There will only be one instance of a teacher employing technology. Since there are numerous technological possibilities available to instructors, it is their responsibility to use them all. The next step in promoting technology in the classroom is to use it.

7.8 DATA COLLECTION

One of the essential elements of mobile learning success is technical competence [30]. For this study, the course centred on computer science and applications for undergraduate degree was chosen. The subjects consists of five assessments with different weight that contribute to the overall subject result, namely CIA-1 (2.5%), CIA-II (2.5%) model (10%) assignment (5%), and a final examination (75%). Table 7.2 depicts the descriptive statistics on academic performance for 3 years, Table 7.3 shows normality test on

Table 7.2 Descriptive statistics on academic performance for three years

Assessment	V-College	N	Mean	Standard deviation (SD)	Standard error (SE)	95% confidence interval for mean		Min.	Max.
						Lower bound (LB)	Upper bound (UB)		
CIA-I	2016	377	64.11	17.371	0.895	62.35	65.87	19	100
	2017	383	63.27	16.933	0.865	61.57	64.97	23	100
	2018	674	52.86	16.499	0.634	51.62	54.11	16	99
	Total	1434	58.60	17.663	0.466	57.68	59.51	16	100
CIA-II	2016	377	87.79	9.582	0.493	86.82	88.76	27	100
	2017	383	91.43	11.048	0.565	90.32	92.54	7	100
	2018	674	86.54	11.363	0.438	85.68	87.40	20	100
	Total	1434	88.18	11.018	0.291	87.61	88.75	7	100
Model	2016	377	74.34	9.716	0.500	73.36	75.33	0.44	96
	2017	383	72.98	10.906	0.557	71.88	74.07	37	96
	2018	674	68.67	9.836	0.379	67.93	69.41	38	95
	Total	1434	71.31	10.409	0.275	70.77	71.85	37	96
Assignment	2016	377	65.09	14.944	0.770	63.58	66.61	25	100
	2017	383	64.25	16.683	0.852	62.57	65.93	15	100
	2018	674	60.12	15.743	0.606	58.93	61.31	10	100
	Total	1434	62.53	15.951	0.421	61.70	63.36	10	100
Examination	2016	377	52.93	17.279	0.890	51.18	54.68	10	99
	2017	383	54.98	18.798	0.961	53.09	56.87	6	99
	2018	674	56.11	19.170	0.738	54.66	57.56	5	100
	Total	1434	54.97	18.623	0.492	54.01	55.94	5	100
Overall	2016	377	63.89	12.906	0.665	62.58	65.19	33	98
	2017	383	64.25	14.106	0.721	62.83	65.67	25	98
	2018	674	62.64	13.768	0.530	61.59	63.68	28	98
	Total	1434	63.39	13.649	0.360	62.69	64.10	25	98

Table 7.3 Assessment scores with skewness and kurtosis for three years (normative test)

Assessment	V-College (N=377)		V-College (N=383)		V-College (N=674)	
	Skewness	Kurtosis	Skewness	Kurtosis	Skewness	Kurtosis
CIA-I	−0.40(0.13)	−0.57(0.25)	−0.20(0.125)	−0.65(0.25)	0.13(0.10)	0.48(0.19)
CIA-II	−2.39(0.13)	9.35(0.25)	−4.57(0.125)	28.59(0.25)	−2.05(0.10)	6.40(0.19)
Model	−0.31(0.13)	−0.25(0.25)	−0.32(0.125)	−0.19(0.25)	−0.02(0.10)	−0.19(0.19)
Assignment	−0.34(0.13)	−0.27(0.25)	−0.32(0.125)	−0.14(0.25)	−0.05(0.10)	−0.30(0.19)
Examination	−0.02(0.13)	−0.55(0.25)	−0.12(0.125)	−0.63(0.25)	0.09(0.10)	−0.66(0.19)
Overall	−0.09(0.13)	−0.53(0.25)	−0.14(0.125)	−0.52(0.25)	0.12(0.10)	−0.57(0.19)

Table 7.4 Academic performance using one-way ANOVA

Assessment	Sum of squares	Difference	Mean square (MS)	F	Sig.
CIA-I	41993.34	2.0	20996.67	74.17	0.00
CIA-II	5914.02	2.0	2957.01	25.18	0.00
Model	9230.42	2.0	4615.21	45.23	0.00
Assignment	7529.02	2.0	3764.51	15.09	0.00
Examination	2445.26	2.0	1222.63	3.54	0.03
Overall	758.86	2.0	379.43	2.04	0.13

assessment scores with skewness and kurtosis for 3 years from 2016 to 2018 and finally Table 7.4 shows statistical comparison on academic performance using one-way ANOVA.

Different classes with similar class sizes of about 48 people each were divided up among the students. Throughout the course of the semester, they had to routinely attend lectures and tutorials. The same set of instructional resources and the same teaching schedule were applied. A ll other assessment components were made available in accordance with the same teaching schedule, with the exception of the test and final exam, which were organised centrally by the college. As a result, each student had the same amount of preparation time and testing time. The evaluations were created to be equally challenging even though they varied between semesters.

7.9 CONCLUSION AND SUGGESTION

The study found that, in view of the advantages it brings, colleges of education may strongly advocate the usage of mobile learning in any social networking application. Additionally, user satisfaction, behavioral intention, and the environment for mobile learning all play a significant role in how well a tool is used. These ultimately lead to a better educational system. Less need for infrastructure like classrooms and books, a better reading culture, and increased student achievement are all signs of an improved educational system. The use of a cellular phone as a teaching methodology has an advantage in education to participate in the teaching-learning process, teaching tool, facilitate the management and plan the teaching process. Even though M-learning can never completely replace traditional learning, when used properly, it may make existing learning methods more valuable.

One of the solutions to the issues facing education is developing as mobile learning. Mobile learning offers more alternatives for the customisation of learning because a variety of tools and resources are constantly accessible. When using mobile learning in the classroom, students frequently collaborate, work in teams, or work alone to complete tasks, fulfil individual needs, and exercise their right to voice and choice. With so much content available at all times and places, there are many chances for formal communication and informal communication takes place in and out of the classroom. Because of their afford ability and app availability, studies have shown that laptops, mobile tablets, iPod touches, and iPads are among the most often used devices for m-learning. The level of social acceptance that mobile learning receives will have a significant impact on its future. On the other hand, users in underdeveloped nations require M-learning to be as portable, accessible, and reasonably priced as users in wealthy nations do. The capacity to make learning mobile and accessible outside of the classroom or the workplace is the very essence of M-learning. Learners who do not have direct access to learning in these locations can now access learning possibilities thanks to wireless and mobile technologies. Many students in underdeveloped nations struggle to access the Internet or to purchase the technology necessary for learning in an online setting.

Utilising mobile learning for academic administration on mobile devices is the first solution. A sizable new revenue stream would be unlocked if it could be proven that mobile learning would take over as the standard way for universities and colleges to alert their student bodies of urgent information. It can be assumed that every student at every university and college has a cell phone that they use frequently. Today, all students enrolled in higher and further education institutions frequently require information from their schools regarding timetable changes, dates for assessments, tutor comments, and other essential administrative details.

The creation and provision of complete modules by mobile learning to students is the third and final tier of the approach for integrating mobile learning into conventional education and training. New technologies now make it possible to create useful course materials. These modules' accessibility, in conjunction with the arrangement of task accommodation, guide interaction, examination, and assessment, will assist in illustrating the reasonability of versatile learning as an engaging source of salary streams for portable suppliers.

7.9.1 Suggestion

With the huge possibility provided by smartphone, our paradigm for education can change. Some of the recommendations are as follows:

1. Strengthen current investments

    The current ICT infrastructure should be supplemented rather than replaced by strategies that are developed by policymakers after taking stock of the current ICT investments and approaches.

2. Localise policies

    When formulating new policies or modifying current ones, policymakers should take into account the local circumstances of the nation or region because approaches that are effective in one nation may not be suitable in another.

3. Back open technological standards

    To enhance access and speed up the development process, policymakers should promote the adoption of open, standards-based platforms for mobile learning applications.

4. Encourage cross-sectoral collaboration and multi-stakeholder alliances

    Government agencies should cooperate with one another, and policymakers should support alliances between various levels and sectors of stakeholders.

5. Create regulations at all levels

    Regardless of whether education is decentralised, policymakers should develop or update mobile learning policies at both the national and local levels. Local policies should drive implementation in specific areas or institutions, while national policies should give overall structure and guidance.

6. Examine and amend current regulations

    The usage of mobile technology in schools and universities may be subject to existing policies that are unduly restrictive. This is especially true of municipal policies. To provide districts and institutions with greater direction, it can be necessary to clarify or amend national policies.

7. Make inclusive education a priority

Policymakers should make sure that mobile learning practises support accessibility for learners with impairments and gender equality. In order to achieve the EFA goals of delivering high-quality education to all learners worldwide, this endeavour is crucial.

REFERENCES

1. Sawsen Lakhal, Hager Khechine, and D. Pascot, "Evaluation of the Effectiveness of Podcasting in Teaching and Learning", *Proceedings, E-Learn: World Conference on E-Learning in Corporate, Government, Healthcare, and Higher Education (E-Learn2007)*, October 15, 2007, ISBN: 978-1-880094-63-1. Association for the Advancement of Computing in Education (AACE), San Diego, CA.
2. Khe Foon Hew," Use of Audio Podcast in K-12 and Higher Education: A Review of Research Topics and Methodologies", *Educational Technology Research and Development*, Vol. 57, No. 3, pp. 333–357, June 2008.
3. Sureshkumar, A., Samson Ravindran, and R. Swarm, Fuzzy Based Cooperative Caching Framework to Optimize Energy Consumption Over Multimedia Wireless Sensor Networks. *Wireless Personal Communications*, Vol. 90, pp. 961–984, 2016. https://doi.org/10.1007/s11277-016-3274-0
4. Douglas Mcconatha, Matt Praul, and Michael J. Lynch,"Mobile Learning in Higher Education: An Empirical Assessment of a New Educational Tool",*The Turkish Online Journal of Educational Technology-TOJET*, Vol. 7, No. 3, July 2008.
5. Sangeeta Paliwal,"Future Trend of Education-Mobile Learning", *International Conference on Academic Libraries*, October 2009.
6. Mohammed Osman M, EI-Husseinand Johannes C. Cronje,"Defining Mobile Learning in the Higher Education Landscape", *Innovationsin Designing Mobile Learning Applications*, Vol. 13, No. 3, pp. 12–21, July 2010.
7. M. Kumaresan, M. J. Basha, P. Manikandan, S. Annamalai, R. Sekaran, and A. S. Kumar, "Stock Price Prediction Model Using LSTM: A Comparative Study," *2023 3rd Asian Conference on Innovation in Technology (ASIANCON)*, Ravet IN, India, pp. 1–5, 2023. doi: 10.1109/ASIANCON58793.2023.10270708
8. Nilgun Ozdamar Keskin, and David Metcalf," The Current Perspectives, Theories and Practices of Mobile Learning", Turkish Online Journal of Educational Technology, Vol. 10, No. 2, pp. 202-208, April 2011.
9. R.K. Dhanaraj, S.K. Islam, and V. Rajasekar, A Cryptographic Paradigm to Detect and Mitigate Blackhole Attack in VANETEnvironments. *Wireless Networks*, 1–16, 2022.
10. Zahra Taleb, and Amir Sohrabi,"Learning on the Move: The use of Mobile Technology to Support Learning for University Students", *Procedia Social and Behavioral Sciences*, Vol. 69, pp. 1102–1109, 2012.
11. Sakeena Fathima and Jarina Begum,"Adaptation of Mobile Learning in Higher Educational Institutions of Saudi Arabia", May 2013.

12. Suresh Kumar Arumugam, Amin Salih Mohammed, Kalpana Nagarajan, Kanagachidambaresan Ramasubramanian, S. B. Goyal, et al., "A Novel Energy Efficient Threshold Based Algorithm for Wireless Body Sensor Network", *Energies*, Vol. 15, No. 16, 6095.

13. Tulika Bansal, and Dr. Dhananjay Joshi," A Study of Student's Experiences of Mobile Learning", *Global Journal of Human Social Science: H Interdisciplinary*, Vol. 14, No. 4, 2014.

14. R. Shankar, B.K. Sarojini, H. Mehraj, A.S. Kumar, R. Neware, A. Singh Bist, Impact of the learning rate and batch size on NOMA system using LSTM-based deep neural network. *The Journal of Defense Modeling and Simulation*, Vol. 20, No. 2, 259–268, 2023. doi:10.1177/15485129211049782

15. Arpit Saxenaand, and Ankit Saxena,"A View point and Attitudes of Students towards Future of Mobile Learning in Education Industry of India", *International Journal of Management, MIT College of Management*, Vol. 3, No.1, pp. 18–22, January 2015.

16. Mostafa Al-Emran, Hatem M. Elsherifand, and Khaled Shaalan," Investigating Attitudes towards the use of Mobile Learning in Higher Education", *Computers in Human Behavior*, Vol. 56, pp. 93–102, March 2016.

17. Kusum Fouzdarand, and Santosh KumarBehera," Attitude of Post Graduate Students towards Mobile Learning", *International Journal for Educational Studies*, Vol. 9, No. 2, February 2017.

18. A. Mehbodniya, A. Suresh Kumar, K.P. Rane, K.K. Bhatia, and B.K. Singh, "Smartphone-Based m Health and Internet of Things for Diabetes Control and Self-Management," *Journal of Health Care Engineering*, Vol. 2021, 2021.

19. Ruey-Ming Chao, Shaio Yan Huang, Jason C.H. Chenand Jia-Nan Chang," Development of STS Collaborative Tutoring Strategy for U-Learning Environment", *International Journal of Mobile Learning and Organisation*, Vol. 3, No. 4, pp. 366–380, July 2009.

20. Maria Virvou, and Eythimios Alepis," Mobile Educational Features in Authoring Tools for Personalized Tutoring", *Computers & Education*, Vol. 44, No. 1, pp. 53–68, January 2005.

21. M. Mallick, M.K.A.S. Govindaraju, A.S. Kumar, M. Kandasamy, and P. Anitha, "Analysis of Panoramic Images using Deep Learning for Dental Disease Identification," *2023 Third International Conference on Artificial Intelligence and Smart Energy (ICAIS)*, Coimbatore, India, pp. 1513–1517, 2023. doi: 10.1109/ICAIS56108.2023.10073939

22. Nadire Cavus, and Huseyin Uzunboylu," Improving Critical Thinking Skills in Mobile Learning", *Procedia-Social and Behavioral Sciences*, Vol. 1, No. 1, pp. 434–438, 2009.

23. www.brainscape.com/blog/2010/09/characteristics-of-effective-mobile-learning

24. Charles Fisher, David C. Dwyer, and Keith Yocam, *Education and Technology: Reflections on Computing in Classrooms*, pp. 336, Wiley, ISBN: 978-0-787-90238-4. June 1996.

25. PeggyA. Ertmer, AnneT. Ottenbreit-Leftwich, Olgun Sadik, Emine Sendurur, and Polat Sendurur," Teacherbeliefs and Technology Integration Practices: A

Critical Relationship", *Computers & Education*, Vol. 59, No. 2, pp. 423–435, September 2012.

26. Jared Keengwe, Grace Onchwari, and Patrick Wachira,"Computer Technology Integration and Student Learning: Barriers and Promise", *Journal of Science Education and Technology*, Vol. 17, No. 6, pp. 560–565, December 2008.

27. Joan Hughes,"The Role of Teacher Knowledge and Learning Experiences in Forming Technology-Integrated Pedagogy", *Journal of Technology and Teacher Education*, Vol. 13, No. 2, April 2005.

28. R. Hermans, J. Tondeur, J. VanBraakand, and M. Valcke,"The Impact of Primary School Teachers' Educational Beliefs on the Classroom use of Computers", *Computers& Education*, Vol. 51, No. 4, pp. 1499–1509, December 2008.

29. Muasaad Alrasheedi, and Luiz Fernando Capretz," Determination of Critical Success Factors Affecting Mobile Learning: A Meta–Analysis Approach",*Turkish Online Journal of Educational Technology*, Vol. 14, No. 2, pp. 41–51, April 2015.

30. Robert D. Hannafinand Wilhelmina, and C. Savenye,"Technology in the Classroom: TheTeacher's New Role and Resistance to it", *Educational Technology*, Vol. 33, No. 6, pp. 26–31, June 1993.

Chapter 8

Pseudorandom bit generator implementation using the dual-CLGG RWT algorithm

S. Bhuvaneswari[1], A. Gayathiri[1], P. Sumitra[1], S. Sabitha[1], G. Sathya[1], and A. Suresh Kumar[2]

[1]PG and Research Department of Computer Science & Applications, Vivekanandha College of Arts and Sciences for Women (Autonomous), Namakkal, Tamilnadu, India
[2]School of Computer Science and Engineering, Jain (Deemed to be University), Bengaluru, Karnataka, India

8.1 INTRODUCTION

A hidden writing technique is called steganography. Since graphics refer to writing, the word steganos indicates covered. Steganography thus encompasses both the art of concealing data and the fact that secret data is being transmitted. Only the recipient is aware that the communication even exits thanks to steganography, which conceals the secret data in another file. In the past, information was safeguarded by hiding it on the scalps of slaves, the backs of wax, writing desks, and rabbits' stomachs. The majority of individuals nowadays, however, send data across the medium in the format of audio, video, pictures, and text. Encryption is used with multimedia assets, such as music, video, and photos, to safely transmit sensitive information, the study of covert communication is known as steganography. Steganography often deals with techniques for concealing the existence of conveyed data so that it can be kept private. It safeguards the privacy of two parties' communications. By incorporating it into the cover picture and creating a stego-image, steganography hides information. Steganography techniques contains so many forms, and each one contains so many merits and demerits. Communication is a crucial component of every expanding business, and today's society commonly uses steganography, utilising different security and data concealment techniques like LSB, ISB, MLSB, etc. Privacy and discretion are ideals shared by all [1].

Accordingly, we can frequently transmit and receive data across secure channels like the phone and the internet, these technologies are not totally secure. Two tactics may be employed to secretly distribute the data. These techniques include things like cryptography and steganography. Cryptography is the process of transforming a communication using a secret key that can only be known by the sender and recipient. Anyone without access to the encryption key cannot read the communication. Even though an encrypted communication can be forcibly intercepted,

DOI: 10.1201/9781003362685-8

assaulted, or decrypted during transmission, steganographic solutions have been developed to address these problems. Steganography, or "as steganography," is the art and science of communicating while concealing the identity of the recipient.

This is how steganography conceals data's presence so that it cannot be recognised. Steganography is the process of hiding informational material inside of any sort of multimedia content, such as a photograph, audio file, or video. The secrecy of data communication can be increased by combining the two ways.

8.2 STEGANOGRAPHY IN IMAGES

Information may now be hidden using images. A photograph with a secret message may be easily sent online or through newsgroups. Niels Provos, a German steganography enthusiast, has studied how steganography is used in newsgroups. He developed a scanning cluster that can search through uploaded photos for hiding messages. One million images were analysed for concealed messages, but nothing was found, proving how limited steganography's real-world uses are. The method of hiding data inside a picture so that it cannot be accessed by unauthorised users is known as steganography, sometimes known as image steganography. This is shown in Figure 8.1.

The cover source can be updated in highly populated regions with a broad variety of color variations to cover a message inside an image without diminishing its aesthetic value or drawing attention to the changes. The least-significant bit (LSB), masking, filtering, and imaging modifications are the most commonly used methods for obscuring these changes. With variable degrees of effectiveness, these methods may be used to different kinds of photo files. The project's goal is to increase public awareness of the many types of steganography that are accessible.

Images are subjected to steganography, and the required information must also be decoded in order to recover the message image. Image steganography is investigated since there are several ways to accomplish this, and one of the techniques is used to illustrate it. Picture steganography

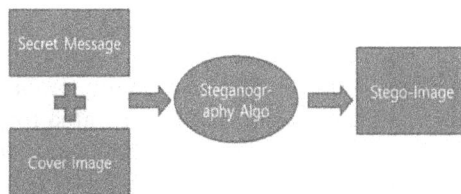

Figure 8.1 Process of steganography.

is the practice of concealing information, such as audio, text, or image files, within another image or video file. Using a spatial domain approach, a picture will be steganographically blended with another image in the current work. Mat lab programmes are used to encrypt and decrypt images, and the only way to obtain this secret information is through an appropriate decoding strategy.

Steganography is concealing information inside other information to hide the fact that communication is occurring. Even if there are other options, digital images are still the most used carrier file type. This is partly because to how common they are online. Using a number of techniques, steganography aims to conceal important information in photographs. They have a broad selection. Each has pros and cons of its own, with some being more advanced than others. The chosen steganography technique must be able to handle the demands of different applications.

For instance, certain applications may need that the secret data be completely undetected, while others may call for the concealment of more important secret information. The hidden message is seen in this image. By enabling the user to select which bits should be swapped rather than merely swapping the LSB from the image, the project offers a safer alternative. The sender chooses a cover picture and inserts the secret text or text file into it via bit substitution over both public and private communication networks. Either on the side receiving it or the side it is opposing. The recipient uses the programme to read the stegno image's hidden text.

8.2.1 Advantages of steganography

1. The key benefit of this approach is that your messages remain secure without being known to third parties.
2. Because the number of bits has been changed based on the user or sender, a third party cannot query the password.
3. The average network users cannot predict the image.
4. Anyone cannot identify a steganographic suspect by glancing at the photographs.
5. It is dependable.

8.2.2 How does random number work?

A random number is one that is chosen equally from among a finite set of possibilities or from an infinite collection. Random number in mathematics and statistics is either statistically random or generated for as a component of a collection that exhibits statistical unpredictictability. Although "1 2 3 4 5 is not random" and "47 88 1 32 41" are commonly accepted statements we cannot state with certainty that the first sequence is not random. It might have come about by accident.

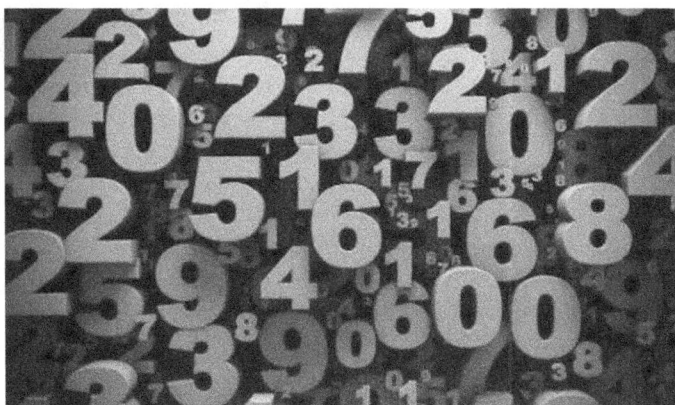

Figure 8.2   Random number.

8.2.3 Pseudorandom number

The only way to produce a really random number is to observe a random physical event, such as dice throws. But when it comes to the computer program or computational machine, which has to generate a random number, than at the back there always will be some mathematical process or a function or a particular algorithm procedure, which will produce a random number. So ultimately here random numbers are not actually random as it follows some sequence and hence it is known as "pseudorandom number." This is shown in Figure 8.2.

8.2.4 Applications of pseudorandom number

Random numbers are used in cryptography, games, and many statistical models.

• For example, construct cryptography keys for data encryption.
• For example, the conduct of a computer controlled character in video games.
• For example, using the Monte Carlo method in statistics.

8.2.5 Linear congruential method (cryptography)

The middle-square technique does not represent the efficient technique due to its potential to repeat or quickly degenerate to zero. The linear congruence method yields a more reliable result. In 1951, Derrick H. Lehmer invented this technique. But since, it has proven to be one of the most prominent PRNGs.

A method called a generator of linear congruence produces a succession of numbers that appear random from a discontinuous piecewise linear equation (LCG). The technique in issue is a tried-and-true technique for generating fictitious random numbers. On computer hardware that implements modular arithmetic using storage-bit truncation, they are very quick and straightforward to implement. Their main concept is also understandable.

This technique uses the following equation:

$$X_{N+1} = (a^* x_n + b) \bmod c \tag{8.2.1}$$

a, b, and c are integer values, and the seed value is $X0$.

Assuming $X_1 = (1*7+7) \bmod (10) = 4$ and $X_0 = 7$, if $a = 1$, $b = 7$, $c = 10$, and $x_0 = 7$, then repeat to create the sequence shown below:

7, 4, 1, 8, 5, 2, 9, 6, 3, 0, 7, 4, 1, 8, 5, 2, 9.... .

$$X_{N+1} = (a^* X_n + b) \bmod c. \tag{8.2.2}$$

The generator may be set to have a complete period of C by selecting the parameters a, b, c, and $x = 0$. These options impact the timescale (the number of terms in a cycle). High C values ensure prolong cycles.

8.2.6 Mersenne twister

It makes use of the Mersenne twister pseudorandom number generator (PRNG). It is undoubtedly the PRNG that is utilised the most in ordinary applications. The Mersenne twister was chosen as the period length when Makoto Matsumoto and Takuji Nishimura invented it in 1997. One cycle would probably take longer to complete than the time it would take for all of humanity to exist in the future. By carefully examining enough of the Mersenne twister's output, one may anticipate all future numbers.

8.3 LITERATURE SURVEY

8.3.1 Production of pseudorandom bits utilising the VLSI architecture of the coupled variable input LCG method

A secure method to produce a pseudorandom bit sequence is to use one of the numerous LFSR, LCG, and chaotic-based PRBG algorithms is the dual-coupled-LCG (dual-CLCG). Due to the inclusion of inequality equations, the hardware implementation of this is sluggish. Using the dual-CLCG technique, direct architectural mapping is initially carried out. Due to the connection of two equations of inequality, the creation of pseudorandom bits is irregular intervals of time leads in a wide range of output delay. It also takes up a lot of room and travels too swiftly. To overcome the various conditions, the "coupled variable input LCG (CVLCG)" technology and related architecture were developed [2].

The linking of two freshly developed variable input LCGs is one of the approaches. And it also contains unique feature. Which are smaller in size in the comparator than dual-CLCG designs and generate pseudorandom bits at regular, steady clock rates. Prototyping is done with a widely accessible FPGA chip, and the architecture is implemented using Verilog-HDL. The sequences are also recorded using a logic analyser, and their randomness is checked using a test tool that is in accordance with NIST standards. The results of the experiments demonstrate that the PRGB approach consistently passes all randomisation tests. To efficiently generate pseudorandom bit sequences for IoT-enabled applications, hardware architecture is being developed. The dual-CLCG approach's hardware architecture is currently being looked at [3].

The unreliable pseudorandom bit generation in this approach causes unexpected output delay and falls short of the maximum time interval. To solve these problems, the "coupled variable-input LCG (CVLCG)" PRBG technology and associated VLSI architecture were developed. The CVLCG approach may yield the longest sequence while simultaneously generating pseudorandom bits at each iteration by connecting two variable-input LCGs. Furthermore, compared to the dual-CLCG configuration now in use, it saves one comparator space [4].

The FPGA implementation's findings demonstrate that the CVLCG architecture can generate pseudorandom bits with just one clock cycle of delay, little hardware, and low power. Findings from the NIST test battery revealed that every difficult test, including those with linear complexity, was consistently passed. Due to its characteristics like reduced space, low latency, and low power, extended period, pseudorandom bit generation at the uniform clock rate, and a high degree of unpredictability, the CVLCG method is suitable for real-time secure information transmission and data security in the lightweight IoT-enabled smart devices [5].

8.3.2 Revision of the dual-CLCG method for a pseudorandom bit generator

With the least amount of VLSI complexity required, the generator for dual-coupled linear congruence (dual-CLCG) has been enhanced to generate pseudorandom bits with high unpredictability characteristics. In this method, a multiplexer circuit with a pick line under the control of the current seed value produces the final random bits. The inequality comparison bits are chosen at random for the multiplexer circuit. In the style of a modified dual-CLCG and its VLSI architecture, this work presents a protection against the more secure random bit generator. They represent an innovative method of using random bit generators. The effects of modifying the architecture of a dual CLCG are observed in order to evaluate some of the critical timing performance characteristics, including beginning clock delay, highest frequency of bit production, and output-to-output [6].

8.3.3 Redundant discrete wavelet transform-based real-time QRS detection

Improvements have been made to generator of dual-coupled linear congruence (dual-CLCG) to create pseudorandom bits with high unpredictability characteristics while using the least amount of the electrocardiogram (ECG) test yields a plethora of knowledge regarding the current state of the heart's health. Processing electrocardiographic data presents a number of challenges, one of which is identifying QRS complexes. A redundant discrete wavelet-based QRS complex detector is displayed in real-time (RDWT). In the approach, wavelet coefficient energy is employed for detection together with scales and wavelet coefficients. The method was tested with the use of the MIT-BIH Arrhythmia Database and demonstrated to be successful in identifying P and T waves, in addition to QRS complexes based on wavelet coefficients and QRS locations [7].

8.3.4 Optimal genetic algorithm-based analysis of the MIMO-RDWT-OFDM system's reliability

Multiple input, multiple output, orthogonal multiplexing of the frequency (MIMO-OFDM) is necessary to gain exceptionally large data rates since the increased need for fifth generation wireless networks with less inter carrier and inter-symbol interference interference (ISI) (UHDR). OFDM based on a discrete wavelet transform (DWT-OFDM) system stands out as a key solution in contrast to conventional OFDM systems because it uses orthogonal wavelets to achieve good orthogonality, giving it more opportunities to lessen the effects of ISI and ICI due to multipath attenuation environments and enhancing the effectiveness of additional bandwidth.

Down sampling increases the true size of the input bit streams while decreasing the efficiency of DWT-OFDM systems. To improve spectral efficiency, instead of using the redundant discrete wavelet transform (R-DWT), discrete wavelet transforms (DWT). A MIMO-RDWT-OFDM system's ability to decrease inter carrier interference (ICI) by reducing jitter noise and improving CIR performance is also investigated in this work. With our approach, the most effective genetic algorithm was used to acquire the correct weights for the ICI cancellation procedure (OGA). The MIMO-RDWT-OFDM system has improved in terms of the bit error rate (BER), speed of processing, and CIR performance [8].

8.3.5 2⁷-1 pseudorandom bit sequence generation 65 nm standard CMO, low power, 11.8 GBPS

When using standard 65 nm CMOS technology, this study shows how a 11.8 Gb of data 27-1 pseudorandom bit sequence (PRBS) generator can manage large amounts of data. The full-rate architecture of the PRBS generator reduces circuit complexity and increases power efficiency. Due to the circuit's D flip-flops (DFF) optimisation is achieved in addition to the asynchronous XOR structure. Low power operation is also feasible. The entire circuit is built using current-mode logic (CML). According to the outcomes of the post-layout simulation, it has a merit value (FOM) of 0.88, requires 73mW from a 1.2V supply, and encompasses 5.5 ps of peak-to-peak jitter.

8.3.6 Watermarking scheme for RDWT domains using a blind and hybrid model

With the use of a hybrid and blind watermarking technology, digital photos are copyrighted. The redundant discrete wavelet transformations (RDWT) and the discrete curvelet transform (DCUT) are the two complicated transforms that are combined in this method (RDWT). These two changes together improve the readability of the watermarking method. The high frequency curvelet coefficients of the cover image are subjected to single level RDWT in order to produce hybrid coefficients that satisfy the system's imperceptibility requirements. By adjusting the wavelet coefficients of the LH sub-band using PN sequences in accordance with the watermark bit and gain factor, the watermark information is introduced.

The method is made more secure by watermarking the image before incorporating it via Arnold scrambling. Numerous real-world images are included in the design's testing. The testing outcomes show the system's superior undetectability and resilience to a variety of attacks compared to more conventional techniques. The technique also performs better than a number of already employed techniques.

8.3.7 For the purpose of improving a quality metric strength factor for an adjustable image watermark, RDWT-SVD and an artificial bee colony were used

Utilising photo watermarking for security can help with authenticity, copyright protection, and other related problems proper data ownership in digital form. The watermark is an image that is binary or grayscale utilised in contemporary watermarking techniques. This offers a brand-new, flexible, and reliable watermarking technique where the host and the watermark are both identically sized, three-dimensional colored images. Arnold's chaotic map, which complicates both the color host and watermark pictures, is used to increase the security of the watermarking technique. The host picture is divided into four sub-bands of the dimension using redundant discrete wavelet transform (RDWT), and the approximate sub-band is then subjected to singular value decomposition (SVD) to produce the principal component (PC). The main component of the watermark is then immediately displayed.

8.3.8 Redundant wavelet transform used in a blind watermarking technique

Another method to safeguard digital photographs' intellectual property is to add a watermark. This paper proposes a hybrid blind watermarking method that combines RDWT and SVD to strike a compromise between imperceptibility and resilience. Using the host image's changed entropy, the watermark's embedding is found. Examination of the orthogonal matrix U generated by the hybrid RDWT-SVD technique is necessary for the use of watermark embedding. To add another layer of security, Arnold's chaotic map is used to mix up the watermark image in binary format. Several geometrical and signal processing attacks are used to test our method. The test results show that our method is more durable and less distorted than other methods when it comes to withstanding JPEG2000 compression, cropping, scaling, and other disruptions.

8.3.9 Dual-CLCG method with pseudorandom bit generation modified for VLSI architecture

In a variety of cryptographic applications, the pseudorandom bit generator is a crucial part for protecting data during transmission and storage (PRBG). The dual-coupled LCG seems to be the most secure of the well-known PRBG systems now in use, including the linear feedback shift register (LFSR), the linear congruential generator (LCG), the connected LCG (CLCG), and the LCG (dual-CLCG). The system's fundamental inequality comparisons produce pseudorandom bits at random intervals. As a result, the twin CLCG method creates a distinctive architecture that produces pseudorandom bits at a constant clock rate. Due to a variety of issues, including a significant beginning clock delay and high memory usage, this architecture makes it challenging to build sequences of the maximum length.

The very large-scale integration (VLSI) architecture of the "modified dual-CLCG" PRBG method attempts to address the aforementioned problems. With a single initial clock delay, little hardware complexity, and a steady clock rate, the PRBG method may produce pseudorandom bits. The PRBG algorithm also completes the 15 benchmark tests in the NIST standard and achieves the maximum period of $2n$. This design was created using Verilog–HDL, and prototyping was carried out on an FPGA chip that is readily available. The updated dual-CLCG technique's 32-bit hardware design is demonstrated to be the optimum choice and may be utilised in performance research that examines hardware complexity, unpredictability, and security. The 32-bit hardware architecture of the updated dual-CLCG method is determined to be the optimum choice and could be helpful for IoT applications and hardware security. The modified dual-CLCG approach is found to be the best option, and its 32-bit hardware design may be useful for IoT applications and hardware security.

8.3.10 The FPGA primes used in a true random number generator architecture are fewer

This chapter uses the jitter and meta stability generated by an FPGA board's primitives as seeds of entropy to show how a true random number generator is built, developed, and explained (TRNG). The TRNG architecture is specially housed on a Xilinx Ultra size XCKU040 FPGA board. Fully digital TRNGs are typically built on FPGA using ring oscillators that contain many look-up-table (LUT) blocks. It does, however, show how a functional FPGA-based TRNG architecture can be created using just one PLL, three on-board primitives, and a few other basic logic elements (e.g., 8 D-type flip-flop, 17). The system's initial synchronisation and post-processing only (UT and two counters).

This method enables the construction of completely configurable VLSI systems by significantly reducing the quantity of FPGA configurable logic blocks (CLB) required, the complexity of the circuitry, and the overall power consumption without sacrificing output bit rate. The TRNG architecture has demonstrated its potential for use in security and cyber security network systems when integrated into Internet-of-Things (IoT) and Industrial-Internet-of-Things (I-IoT) applications. The Anderson-Darling, Kolmogorov-Smirnov, National Institute of Standards and Technology (NIST), and statistical aspects of the created 100 Mbps output bit stream tests were successful for the TRNG architecture.

8.4 METHOD OF DUAL-CLCG AND ITS ARCHITECTURE

The following mathematical description demonstrates how the dual coupled linear congruential generator (CLCG) technology connects the four linear congruential generations to generate pseudorandom bit sequences:

$$X_{i+1} = a_1 \times x_i + b_1 \bmod 2n \qquad (8.4.1)$$

$$Y_{i+1} = a_2 \times y_i + b_2 \bmod 2n \qquad (8.4.2)$$

$$P_{i+1} = a_3 \times p_i + b_3 \bmod 2n \qquad (8.4.3)$$

$$Q_{i+1} = a_4 \times q_i + b_4 \bmod 2n \qquad (8.4.4)$$

The following inequality equation can be used to calculate a pseudo-random bit sequence:

$$z_i = \begin{cases} 1 \text{ if } x_{i+1} > y_{i+1} \text{ and } p_{i+1} > q_{i+1} \\ 0 \text{ if } x_{i+1} < y_{i+1} \text{ and } p_{i+1} < q_{i+1} \end{cases}$$

$$(8.4.5)$$

A different method of computing the output sequence Z_i is as follows:

$$Z_i = B_i \qquad \text{if} \qquad C_i = 0$$

$$\text{where, } B_i = \begin{cases} 1, & \text{if } x_{i+1} > y_{i+1} \\ 0, & \text{else} \end{cases} \quad ; \quad C_i = \begin{cases} 1, & \text{if } p_{i+1} > q_{i+1} \\ 0, & \text{else} \end{cases} \qquad (8.4.6)$$

Here, the starting seeds are x_0, y_0, p_0, and q_0, while the steady factor are a_1, b_1, a_2, b_2, a_3, b_3, and a_4 and b_4.

The following requirements must be met to obtain the maximum duration period:

1. b_1, b_2, b_3, and b_4 should be relatively prime with $2n$;
2. Each and every one of the components (a_{1-1}), (a_{2-1}), (a_{3-1}), and (a_{4-1}) must divide by 4.

According to equations (8.4.1), (8.4.2), (8.4.3), and (8.4.4), the linear congruential generators (LCGs) produce the random outputs Q_i+1, X_i+1, Y_i+1, P_i+1, respectively. These outputs and the inequality equation are then used to construct the pseudorandom bit sequence (8.3.5). In addition, it compares X_i+1 to Y_i+1 and P_i+1 to Q_i+1. The result of the dual-CLCG technique is "1" when $X_i+1>Y_i+1$ and $P_i+1>Q_i+1$. In a similar manner, it produce "0" when $X_i+1 < Y_i+1$, and $P_i+1< Q_i+1$. However, under some unfair circumstances, neither a legal output of "1" nor "0" is produced.

8.4.1 Dual-CLCG method is used for architecture mapping

This section thoroughly examines the architectural mapping of the dual-CLCG approach, which is seen in Figure 8.3. A one-bit tristate buffer, two comparators, and four LCG blocks make up its architecture. To produce a-bit pseudorandom binary integer, the dual-CLCG design's LCG, the main component, also requires an adder and a multiplier. When = (2 +1) is taken into account, the shift operation may be used to multiply x in the LCG equation (8.2.1). In this case, this particular positive integer is more than "1" and less than 2. Consequently, equation (8.2.1) might be rewritten as, to calculate

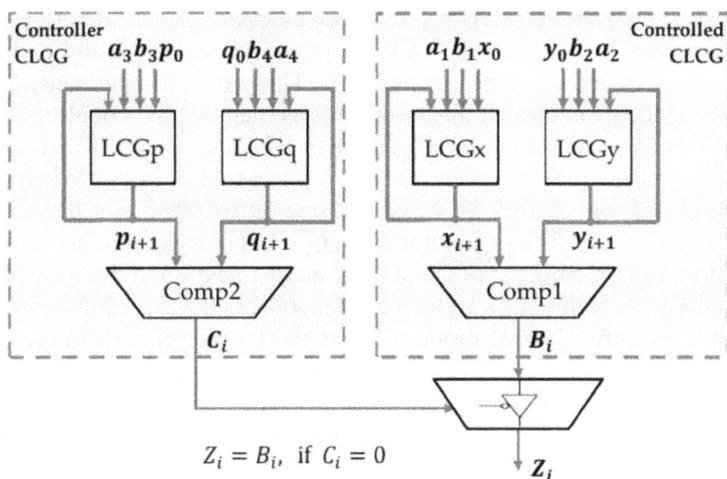

$Z_i = B_i$, if $C_i = 0$

Figure 8.3   Direct architectural mapping of the dual-CLCG method.

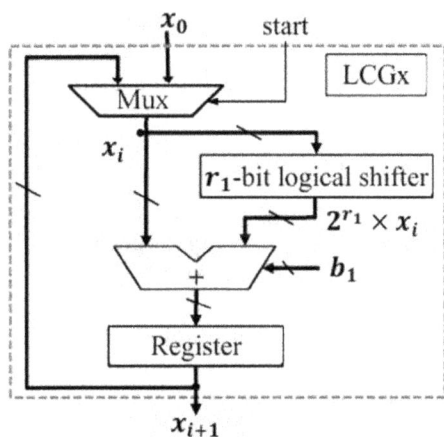

Figure 8.4   The linear congruential generator's architecture.

$X_i+1 = (a_1 \times x_i + b_1) \bmod 2^n = [(2^{r_1}+1)\, x_i + b_1]$
$\bmod\ 2^n = [(2^{r_1} \times x_i) + x_i + b_1)] \bmod 2^n$ 　　　　(8.4.7)

The three-operand modulo $2n$ adder creates the architectural mapping of LCG, as shown in equation (8.4.7) and Figure 8.4. It produces the n-bit random binary output X_i+1 from the original seed x_0 in a single clock cycle. The accompanying design of the LCG equation may also be used to map the three further LCG equations (8.3.2), (8.3.3), and (8.3.4), which compute y_i+1, e_i+1, and q_i+1, respectively, in one clock cycle (8.3.1), (8.3.4), (8.3.4). Equation (8.3.7) compares the LCG outputs x_i+1 with y_i+1 and e_i+1 with q_i+1 using two n-bit binary comparators. These two comparisons, B_i and C_i, provide identical one-bit binary outputs as their results. Equation (8.3.6) also builds two one-bit random "controller-CLCGs," one of which creates C_i and the other of which generates B_i, utilising a tristate buffer. When $C_i = 0$ (the output of the controlled-CLCG), the one-bit tristate buffer picks B_i, the controlled-output, CLCGs, but not when $C_i = 1$. The tristate buffer is chosen at random since the output of CLCG is likewise unexpected. Because of this, the dual-CLCG technique's architectural layout, as illustrated in Figure 8.3, is unable to output a real random bit at the tristate buffer's output once every regular clock cycle. The output latency (output to output delay) in this instance is determined by the value inserted between two 0s in the C_i series of consecutive 1s. There is just one asynchronous PRBG architecture that works.

8.4.2 Encoding process

The header of the image displays the offset. To maintain the integrity of the header, that offset is left in place, and we begin the encoding process with the subsequent byte. We first collect the input carrier file, an image file, and then encode it. The user is then directed to a specific text file after that.

8.4.3 The process of decoding

The offset is contained in the image's header. To generate the user space, use the same steps as for the encoding, utilising the get Raster () and getDataBuffer () methods of the Writable Raster and Data Buffer Byte classes. The picture data is stored in a byte array. The byte array indicated above is used to get the bit stream from the original text file and place it into another byte array. The original message, which included the aforementioned byte array, is also copied to the decoded text file.

8.4.4 Procedure of image data hiding

Step 1: Extract every pixel from the provided image and place it in an array called (image array).

Step 2: Extract each character from the message file and place it in the array referred to as (message array).

Step 3: Put the characters you took out of the Stego key in the Key array. A stego-key controls the hiding process, which restricts access to the embedded data's identification and/or retrieval.

Step 4: The initial pixel as well as the characters from the Key array should be present in the pixel's first component. Add more characters to the Key array by inserting them into the first component of the following pixels.

Step 5: Put a concluding sign there to mark the key's conclusion. In this procedure, the terminal symbol is set to 0.

Step 6: Replace each component of the subsequent pixels with a character from the message array.

Step 7: Until all the characters have been implanted, repeat step 6 as necessary.

Step 8: To mark the end of the data, add another terminating sign.

Step 9: All of the characters entered will be hidden in the produced image.

The simplest steganography techniques just insert. The cover image contains the least important part of the message ordered arrangement of the area. Since the shift's amplitude is so small, modulating the least significant bit does not produce a change that humans can see. If you wish to hide a concealed message inside of an image, you need a compelling cover image. Since this method makes use of bits from every pixel in the image, a lossy compression scheme cannot be employed since the hidden information would be lost during the transformations. In a 24-bit color image, the red, green, and blue color components may each be represented by one bit, giving each pixel a total of three bits to store information.

(00100111 11101001 11001000)

(00100111 11001000 11101001)

(11001000 00100111 11101001)

The grid below shows what is produced when the letter A, whose binary value is 10000001, is used:

(00100111 11101000 11001000)

(00100110 11001000 11101000)

(11001000 00100111 11101001)

The character may appear in the right place in this case by modifying simply three bits. There are times when it is necessary to only change 50% of a picture's elements while utilising the largest cover size to hide a concealed message. The least significant bit changes are too small for the human visual system (HVS) to detect, which allows for the effective concealment of the information. The tiniest bit of the third color, as you can see, has remained unchanged. It may be used to verify the precision of the 8 bits included in these three pixels. Therefore, it may serve as a parity bit.

8.4.5 Dual-CLCG algorithm

In the dual-coupled linear congruential generator (dual-CLCG) method, the four linear congruential generators are coupled to produce pseudorandom bits as follows:

$$X_{i+1} = a_1 \times x_i + b_1 \bmod 2^n \tag{8.4.8}$$

$$Y_{i+1} = a_2 \times y_i + b_2 \bmod 2^n \tag{8.4.9}$$

$$P_{i+1} = a_3 \times p_i + b_3 \bmod 2^n \tag{8.4.10}$$

$$Q_{i+1} = a_4 \times q_i + b_4 \bmod 2^n \tag{8.4.11}$$

The following inequality equation may be used to calculate the pseudo- and ombit sequence.

$$z_i = \begin{cases} 1 & \text{if } x_{i+1} > y_{i+1} \text{ and } p_{i+1} > q_{i+1} \\ 0 & \text{if } x_{i+1} < y_{i+1} \text{ and } p_{i+1} < q_{i+1} \end{cases} \tag{8.4.12}$$

The following equations can also be used to determine the output sequence Z_i.

$$Z_i = B_i \qquad \text{if} \qquad C_i = 0$$

$$\text{where, } B_i = \begin{cases} 1, & \text{if } x_{i+1} > y_{i+1} \\ 0, & \text{else} \end{cases} \quad ; \quad C_i = \begin{cases} 1, & \text{if } p_{i+1} > q_{i+1} \\ 0, & \text{else} \end{cases} \tag{8.4.13}$$

Here, the beginning parameters are x_0, y_0, p_0, and q_0, while the constant parameters are a_1, b_1, a_2, b_2, a_3, and a_4 and b_4. To be eligible for the maximum duration term, the following conditions must be satisfied:

b_1, b_2, b_3, and b_4 should be relatively prime with $2n$;

The integer must have a_{1-1}, a_{2-1}, a_{3-1}, and a_{4-1} sums that are all divisible by 4.

The random outputs X_{i+1}, Y_{i+1}, P_{i+1}, and Q_{i+1} are generated using the linear congruential generators (LCG) denoted by equations (8.4.8), (8.4.9), and (8.4.10), respectively. These outputs and the inequality equation are then used to calculate the pseudorandom bit sequence (8.4.10). It simultaneously compares X_{i+1} with Y_{i+1} and P_{i+1} with Q_{i+1}. When $X_{i+1} > Y_{i+1}$ and $P_{i+1} > Q_{i+1}$, the dual-CLCG technique yields the value "1." In a similar manner, it produces "0" when $X_{i+1} < Y_{i+1}$ and $P_{i+1} < Q_{i+1}$. However, other inequality conditions prohibit it from generating the right answer, such as "1" or "0." The result of the dual-CLCG technique instead selects the value when it is "0," skips the value when it is "1," and does not provide any meaningful binary digits. To construct the random bit sequence, a similar equation may be employed (8.4.6).

8.4.6 Flow diagram of the current system

Figure 8.5 shows the flow diagram of the proposed system. Here pseudorandom bit is generated by the dual-CLCG algorithm and then it is hidden in the image to improve the security and redundancy.

The pseudorandom bit is generated using equation (8.4.5) and then it is converted to data by bits to data encryption. Image is given as input. Here the converted data is hidden in the image with the help of RWT algorithm. RWT algorithm is a routing algorithm in which it is used to locate the exact place of data. Finally the encrypted message and image will be obtained at the end of encryption process. At decryption the key such as public key and private key is given and the encoded image is given as input to decryption panel. Here finally the original image and data which is given as input is gained at the decryption panel.

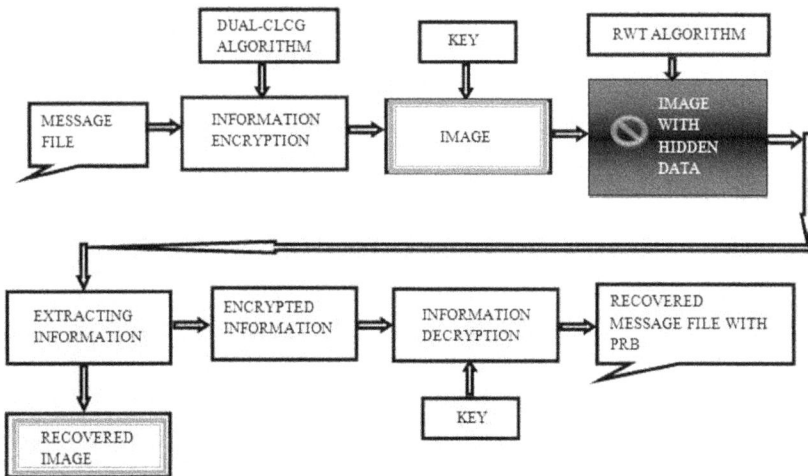

Figure 8.5 Flow diagram of the current system.

8.4.7 Outline of the output panel

From Figure 8.6 the encryption panel and decryption panel is appeared when the code starts run.

8.4.8 Encryption panel input

From Figure 8.7 the inputs such as key values (i.e., ENTERP and ENTERQ), input dimension and input image is given as input to the encryption panel.

8.4.9 Encryption panel output

From Figure 8.8 after the inputs were given, the encryption process will be started after the encryption process completes the output such as public key, private key, encrypted message and encrypted image will be obtained.

8.4.10 Encrypted image

From Figure 8.9 it is shown that the output of encrypted image after encryption process completed.

8.4.11 Decryption panel input

From Figure 8.10 the input such as private key, public key, encoded image and encoded message is added in decryption panel.

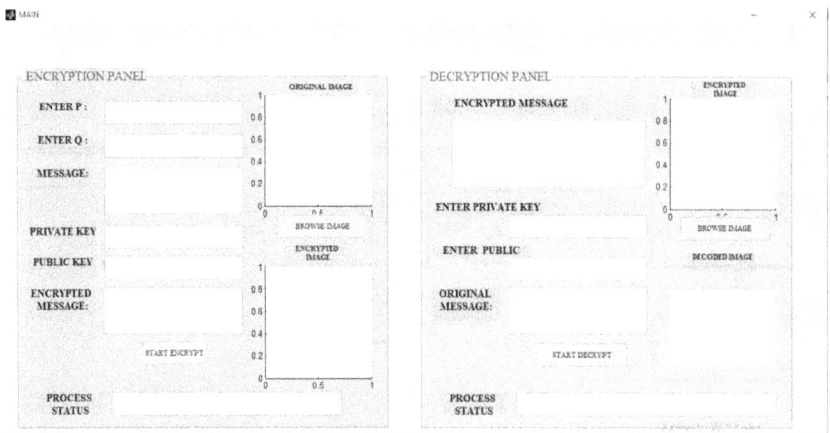

Figure 8.6   Outline of the output panel.

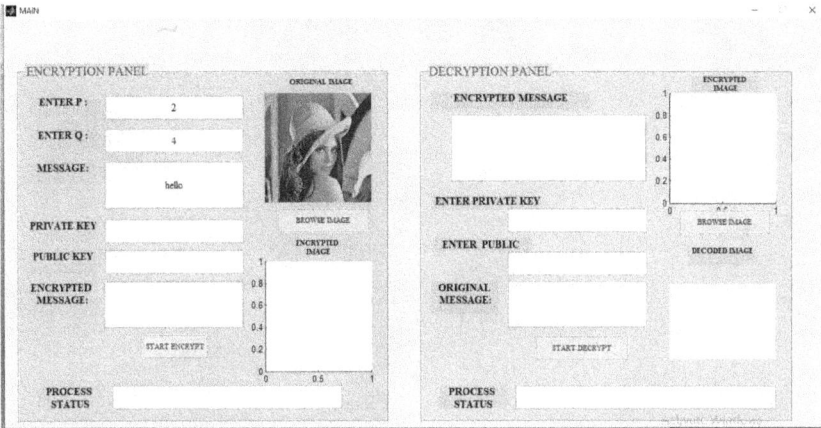

Figure 8.7   Encryption panel input.

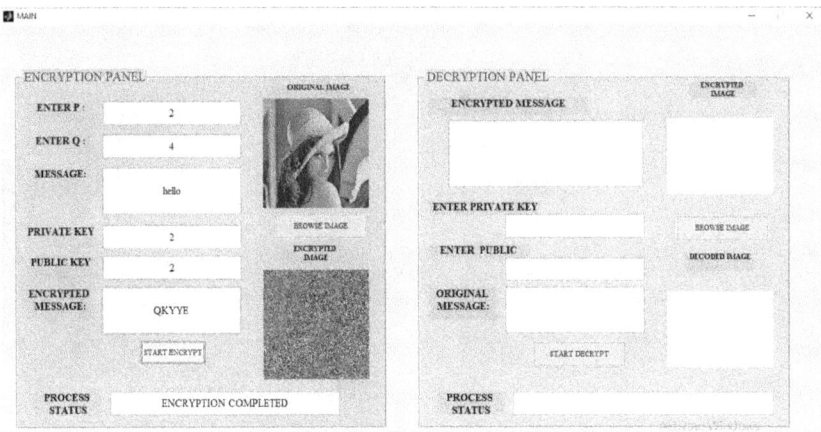

Figure 8.8   Encryption.

Figure 8.9   Encrypted image.

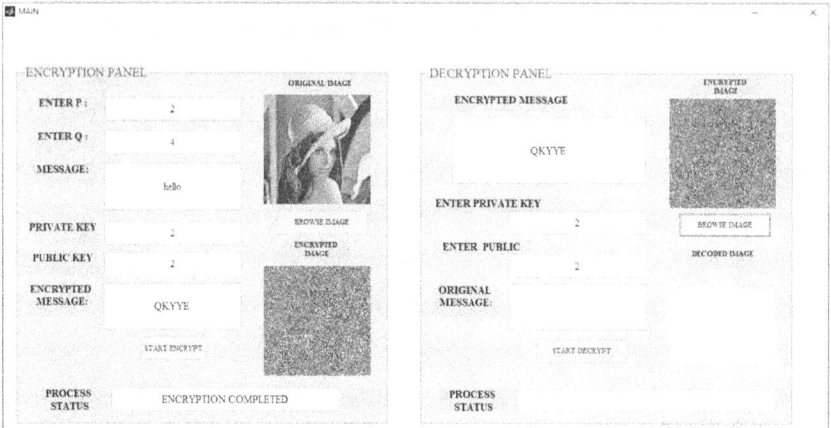

Figure 8.10   Decryption panel input.

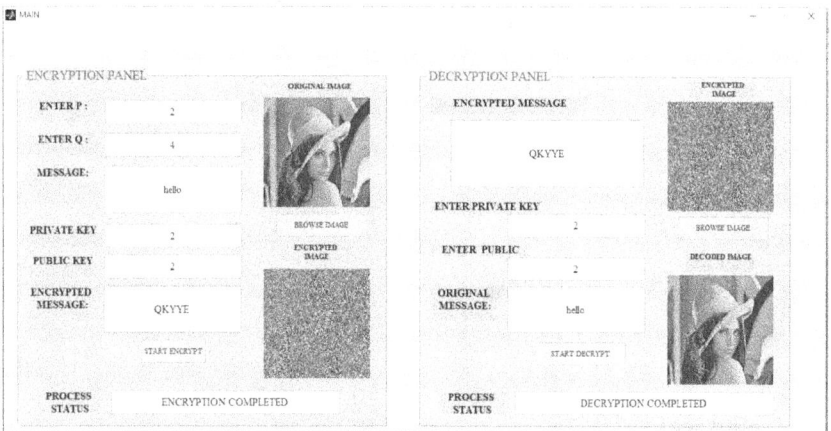

Figure 8.11   Decryption panel output.

8.4.12 Decryption panel output

From Figure 8.11 the final output of the decryption panel is visible. Here the original image and message which is given as input is finally obtained in the decryption panel.

8.4.13 Decrypted image

From Figure 8.12 it shows the decrypted image after the process gets completes. The decrypted image obtained is what we have given as the original image.

Figure 8.12   Decrypted image.

8.5 CONCLUSION

The dual–CLCG technique, which is used to create pseudorandom bits, is applied in this research to increase security and redundancy. Hiding visual data uses the random bit the-dual-CLCG method. The message picture that is provided at the sender side is obtained without any data loss after the image; data and key are first provided as the input. The message image is then encrypted before being delivered together with the encrypted message, image, and keys to the decrypted panel. Here the data is concealed within the image to increase security when sending a data, which enhances communication security.

REFERENCES

1. Amit Kumar Panda, and Kailash Chandra Ray published, "A Coupled Variable Input LCG Method and its VLSI Architecture for Pseudorandom Bit Generation", In *IEEE Transactions on Instrumentation and Measurement*, Volume-69, Issue 4, 2020.
2. R.K. Chauhan, and Deep Gupta Mangal publishes, " Design of Modified Dual-CLCG Technique for Pseudo-Random Bit Generator", In *IEEE Transaction on Very Large Scale Integration (VLSI) Systems*, Volume 28, Issue 7, July 2020.
3. E. Arrais, R. A. M. Valentim, and G. B. Brando," Real Time QRS Detection Based on Redundant Discrete Wavelet Transform", *IEEE Latin America Transactions*, Volume 18, Issue 7, April 2020.
4. Abdul Hussain Sharief, and M. SatyaSairam, "Performance Analysis of MIMO-RDWT-OFDM System with Optimum Evolutionary Algorithm," *International Journal of Electronics and Communication*, Elsevier, Volume 111, November 2019.

5. Y. Wang, M.-D. Wei, and R. Negra, "Low Power,11.8 Gbps 27-1 Pseudo-Random Bit Sequence Generator in 65 nm Standard CMOS." *2019 26th IEEE International Conference on Electronics, Circuits and Systems (ICECS)*, Genoa, Italy, 2019, pp. 318–321, doi: 10.1109/ICECS46596.2019.8965019.

6. Rohit Thanki, Ashish Kothari, and Deven Trived, *"Hybrid and Blind Watermarking Approach in DCuT–RDWT Domain"*, Elsevier, Volume 46, June 2019.

7. Sourabh Sharma, Harish Sharma, and Janki BallabhSharma, "Hybrid and Blind Watermarking Approach in DCuT–RDWT Domain," *Journal of Information Security and Applications–"An Adaptive Colour Picture Watermarking Using RDWT-SVD and Artificial Bee Colony Based Quality Metric Strength Factor Optimization"*, *Applied Soft Computing Journal-Elsevier*, Volume 84, 7 August 2019.

8. Ferda Emawan, and Muhammad Nomani Kabir, "Ablind Watermarking Techniqueusing Redundant Wavelet Transform," *IEEE Transactions on Signal Processing and Its Applications*, Volume 60, May 2019.

Chapter 9

Steganography techniques securely communicate using cryptography

S. Sabitha, P. Sumitra, G. Sathya, M. Sathiya, A. Gayathiri, and S. Bhuvaneswari

PG and Research Department of Computer Science & Applications, Vivekanandha College of Arts and Sciences for Women (Autonomous), Namakkal, Tamilnadu, India

9.1 INTRODUCTION

The art of transforming data into secret writing is known as cryptography. Writing is done using the words cryptography, which meaning secret or hidden. Cipher is a straightforward encryption method that is initially used to secure military communication. Let's look at a straightforward example of message encryption where the succeeding two letters are utilised in place of the actual alphabet. As an illustration, if the data is sent as "data," the recipient will get it as "fcvc". Hidden code needs to decrypt the information, which only the receiver is aware of. If an unauthorised individual received the communication, he will never comprehend the true meaning of what was sent. Numerous cryptographic algorithms existed [1].

9.1.1 Features of cryptography

A crucial component of safe communication is cryptography. There are various criteria specific to cryptography. They are authentication, non-repudiation, data integrity, and data confidentiality.

1. **Confidentiality:** Only the authorised person has access to all information; everyone is not prohibited. By encoding information, its meaning is concealed. Source machine use confidential key to secret information. The message is decrypted by the receiver using a cryptographic key that could or might not match the one used by the sender.
2. **Integrity:** No information can be modified without the recipient's knowledge between a sender and their intended receiver. Integrity uses hashing to build a unique message digest from the message that is forwarded with the message, ensuring that the message sent and received are identical. A second digest is produced from the message using the same

DOI: 10.1201/9781003362685-9

169

method by the recipient to be compared to the first. This method just prevents accidental message modification. To defend against malicious modification, a variant is utilised to produce digital signatures.

3. **Non-repudiation:** Person who creates the information cannot later decide against it. The declaration that no one can dispute anything is termed as non-repudiation. A service that offers evidence of the source and integrity of data is known as non-repudiation and it is a legal notion that is commonly required in information security. It makes it incredibly difficult to convincingly dispute the origins of a communication, its legitimacy, and its integrity. Digital signatures were capable of providing non-repudiation for online transactions.

4. **Authentication:** The letter's sender and recipient are both known. Both the origin of the information and its intended use are verified. Individually person or organisation could be establishing recognition. Achieved with the use of digital certificates. A widely used cryptographic authentication mechanism is Kerberos.

9.2 CRYPTOGRAPHY TYPES

Cryptography comes in three different flavours:

9.2.1 Symmetric key

The encryption method known as cryptography allows the sender and recipient to both encode and decode communications using a synchronised code utilising the same shared key. Symmetric key systems (SKS) are quicker and easier, but confidential code interchange among originator and recipient is necessary. Data encryption standard (DES) popularly used in symmetric key encryption algorithm. With this technique, the secret key was known by both sender and recipient. DES is the most prevalent algorithm [2].

Symmetric key encryption techniques fall into two main categories: stream ciphers and block ciphers.

A feedback mechanism is used by stream ciphers to operate on a single bit by changing code repeatedly.

A method to encrypt data in the form of blocks called block ciphers whereas encrypt binary value at the time called stream cipher.

9.2.2 Hash functions

No need for any special key to rebuild the content. Difficult to rebuild a portion of original information since fixed length hash value is produced based on original information. Hashing techniques using operating system (OS) to maintain confidential key.

9.2.3 Asymmetric key

It is possible to encrypt and decrypt using the keys in a message. Public key to encrypt information and private key to decrypt information.

9.3 TOOLS FOR CRYPTOGRAPHY

The usage of tools is more advantageous for other cryptographic activities like code signing and signature confirmation. The following is a list of the most popular cryptographic tools [3].

9.3.1 Security token

The use of this token verifies the user's identity. A security token should be encrypted to conduct a secure data exchange. The whole state is provided to the HTTP protocol. The state is continued by a browser using the token that was created on the server. Remote authentication is frequently used.

9.3.2 JCA

The encryption method is verified using this tool. The Java library has several functions but you must import them before using them. Even though it's a Java library, it works well with many different frameworks and makes it easier to create plenty of apps.

9.3.3 SignTool.exe

The primary features for adding the sign and interval stamp to any form of organiser are support. Increased file dependability is guaranteed by this signTool.exe executable file with whole feature set.

9.3.4 Docker

Docker allows users to build sizable apps. The whole database that is stored in encrypted manner. Crypto should be carefully followed in order to continue with data encryption in this. Encrypting data and files makes it impossible for anybody to access them without the right access key. By allowing users to manage it on a server, it is portrayed as being in the cloud.

9.3.5 CertMgr.exe

Here, the file is displayed in executable format. Various certificates of CertMgrare are dependable. In addition, it maintains CRLs, which are lists of revoked certificates. The purposes of cryptography in the generation of certificates raise the security and make it possible to include additional layers of security.

9.3.6 Authentication using key

To decrypt information that has been encrypted, keys are needed. Only the intended user may access encrypted information, although everyone can interpret conventional information. Every safe activity makes use of cryptography tools, and there are a variety of them accessible so that customers select based on demands.

9.4 ALGORITHMS FOR CRYPTO

Crypto includes the following algorithms:

Security is the most important factor in this IoT industry. Although there are many security measures in place, they are not able to create modern smart apps, especially for software running on hardware with resource limitations. As a result, cryptography techniques were developed, ensuring increased security [4].

The following are a few examples of cryptography algorithms:

9.4.1 Triple DES

The triple data encryption standard algorithm is used in security techniques in place of the traditional DES mechanism. It enables cyperpunk to eventually acquire information needed to successfully circumvent. This was the strategy that many businesses had widely adopted. Three keys, each with 56 bits, are used in the operation of triple DES. The maximum key length is bits; however experts argue that a key intensity of 112 bits is most likely. This method is capable of producing a trustworthy hardware encrypt solution of institutions and businesses [5].

9.4.2 Blowfish

Blowfish was primarily developed to replace the triple DES methods. It divides data into a 64-bit block each and independently encrypts the search clock. The speed and effectiveness of Blowfish are its most alluring qualities. Since this is an open algorithm that anyone can use, many people benefited from its implementation. This algorithm is used in every area of industry because it has many characteristics for password protection. These factors all contribute to this algorithm's market dominance [6].

9.4.3 RSA

Data sent over the internet is protected using one of the public-key encryption methods. The PGP and GPG systems regularly used this algorithm. RSA is considered a symmetric type of algorithm because it functions with a limited number of keys. An encryption process uses one key, and a decryption process uses a second key [7].

9.4.4 Twofish

In order to guarantee security, this algorithm uses keys, yet only one key is necessary because it employs the symmetric approach. A maximum of 256 bits can be used in the keys of this technique. It is renowned for faster and suitability for applications. It is one of the most extensively used algorithms. Furthermore, it is an openly available, extensively used algorithm [8].

9.4.5 AES (advanced encryption standard)

The US government and many other businesses use this algorithm technique because it is the most reliable. Even while technique works well in 128-bit encryption form, the most common bits used for massive encryption operations are 192 and 256. The AES approach is widely praised for encrypting data in the private domain because it is so impervious to all hacking systems [9].

9.4.6 Advantages

1. AES is very flexible.
2. Compared to other ciphers, security is high.
3. In reality cost is low.

9.4.7 LSB algorithm for steganography

In this chapter, for steganography, the least significant bit algorithm is used. Computer understands binary data. All the high level languages are converted into binary digits. In binary data contains either 0 or 1. For example 00010000, 00011000, 00010001 are the binary data. Here the LSB is the right side last digit. We take cover image as rose flower. Now we implement the steganography means that is here we hide the data with help of LSB. The binary data is changed like 00010001, 00011000, 00010001. In this way we have hidden the data within the image.

9.4.8 Advantages

1. LSB provide security.

2. Flexibility of use.
3. Easy to implement.

9.5 APPLICATIONS OF CRYPTOGRAPHY

Only security-related applications of cryptography were used in practice. To ensure the accuracy and dependability of the transmitter used additional security. Security also became increasingly important with the introduction of digital communications, and as a result, the use of cryptography procedures for ensuring absolute secrecy began to outpace their use.
Here are a few examples of how cryptography is used.

9.5.1 Confidentiality in storage

By keeping encrypted data, cryptography enables users to avoid the main vulnerability that hackers can exploit.

9.5.2 Reliability in transmission

A common technique that encourages dependability is to perform a cyclic redundancy check (CRC) of the transferred data and attached corresponding cyclic redundancy check in scrambled design. CRC with secret information after receiving CRC and secret information one more time check the information. Effective cryptographic techniques are thus considerably more necessary for reliable quarantine message transmission [10].

9.5.3 Identity verification

Innovative systems presumably utilise robust cryptographic techniques along with people's physical techniques and communal secrets to provide extremely reliable identity verification. Cryptography is closely related to the way of using passwords.

9.5.4 Example applications

1. Mobile pin.
2. Encrypted signatures.
3. Secure network browsing.
4. Electronic money.
5. Validation.
6. Crypto currencies.

Plain text is another name for the initial message, and cipher text is another name for the message that results from converting secret data. Encryption is the process of turning a message's original letter into cryptograph letter (secret code). Decryption is the method of converting a secret message back into plain text. Secret keys are necessary for encryption and decryption.

Data concealment is the art of steganography. This is made possible by concealing information in other data. Steganography uses a variety of approaches in several fields to conceal data. The frequency domain uses a variety of converting techniques, such as DCT and FFT, making it extremely difficult to uncover hidden messages in this domain.

9.6 STEGANOGRAPHY IN INFORMATION SECURITY

Steganography is a technique for concealing secret information inside of other messages. This outcome serves as the cover for the hidden message. Audio, video, and image files can all be encrypted using steganography. Hash marks and other symbols are often used to communicate steganography, but it is also frequently utilised within photographs. In any case, steganography makes unlawful reading easier and reduces material theft [11].

Steganography is used in watermarking, which hides copyright information in a watermark by superimposing documents that are invisible to the unaided eye. By doing this, fraud is avoided, and copyrighted media is better protected.

To circumvent generating concerns about the transfer for sensitive information, steganography's primary goal is to maintain safe communication while being completely untraceable. While its purpose isn't to prevent individuals from learning the secret knowledge, it can prevent them from accepting such knowledge. If steganography questions the channel for the transporter, it fails.

Steganography's fundamental model consists of a password, a message, and a carrier. A carrier is used to implant the message and conceal its existence (cover-object). A sender wants the content of the information to still be private. Anything can be installed in a binary stream, such as original message, secret code, another image, a copyright mark, a covert message.

A stego-key is a term for a PIN. It may stipulate address from a cover object can only be extracted by a receiver who is familiar with the corresponding decoding key. The cover-object is then referred to as the stego-object and has the message placed covertly.

Cover-object itself and the related decoding key were needed to retrieve the address from a stego-object. Stego-key utilised through the scrambling procedure. The actual picture needed a majority of programmes to extract the information.

Varieties of suitable carriers are as follows:

- Internet protocols.
- Encoding audio.

- A disc uses free space to append or hide documents.
- Text that solely contains null characters, similar to morse code, such as java and html.
- Image files that can be both color and greyscale.
- The information concealing process often removes unnecessary parts from the cover item.

The procedure consists of the following:

- The unused bits are found in a cover-object. Duplication of bits is modified without affecting cover-object.
- The embedding method selects the portion of unnecessary binary code that should be restored.

9.7 TEXT STEGANOGRAPHY IN INFORMATION SECURITY

It is a technique for covering a secret code as a covering message within another word or by constructing a cover data that is related to the actual secret code. It involves different kinds of methods such as altering text, altering the words themselves and so on. It lacks the duplicate information such as image, audio, or video files. By observing changes to the document's structure, it can conceal information without obviously altering the relevant output [12].

An image or an audio file can have insignificant changes made to them, but a casual reader may notice an extra letter or a missed punctuation mark in a text file. It can save text files, which uses less memory, and is faster and easier to communicate with than other steganographic techniques.
Text steganography can be broadly categorised into three types:

i.   Format-based methods.
ii. Random and statistical production.
iii. Linguistic approaches.

9.7.1 Format-based methods

This technique includes physically altering the text formatting to hidden the contents. This approach has particular drawbacks. The stego file can be opened in a word processor to identify typos and additional white spaces.

A human reader may become suspicious of altered font sizes. Additionally, if the original text is available, comparing it to the alleged steganographic letter can make modified letter elements fairly obvious.

9.7.2 Random and statistical generation

It is possible to prohibit correspondence with a known plaintext in random and statistical generation; steganographers therefore turn to developing their own cover texts. One technique is to hide data by displaying characters in a random order.

Another technique involves creating words that will naturally occur to have similar statistical attributes to existing words in the given language by using statistical features of word length and letter frequencies.

9.7.3 Linguistic steganography

In some circumstances, linguistic steganography uses linguistic mechanisms as the area in which messages are hidden, paying close attention to the linguistic features of created and edited text. CFG used to hidden the binary code when left division of corresponds to "0" and the right division to "1."

It is also possible to apply a grammar in GNF when a production's first choice defines bit 0 and its second choice defines bit 1. There are some drawbacks to this approach. First of all, poor grammar will result in numerous text repeats. Second, despite the text's perfect grammar, semantic architecture has a disadvantage. As a result, the sentences form a random string with no connection to one another.

9.8 IMAGE STEGANOGRAPHY IN INFORMATION SECURITY

The most well-known cover objects in steganography are images. In order to hide the information, pixel intensities are employed in image steganography. Different picture organiser arrangements are in the world of cypernated photos [13].

A photograph is a group of integers representing distinct light forces in various regions of the picture. Numerical descriptions are mentioned as pixels, and they form a grid. The majority of Internet photos include a rectangle representation of the image's pixels (specified as bits), showing the location of each pixel as well as its color. Row-by-row pixels are displayed straight.

The quantity of binary digits used in color design is referred to as bit depth. It can specify how many bits are used for each pixel. Smallest binary digit depth in contemporary color design is 1 byte, meaning that 1 byte is used to specify each pixel's color.

Grayscale and monochrome images can show 16*16 RGB different colors or shades of grey and require 1 byte for each pixel. The RGB color model, often known as a real color, is required for digital color photographs that are typically preserved in 3-byte formats.

Three primary colors, such as red, green, and blue, are modified for every pixel in a 3-byte image; each primary color has a 1-byte definition. Therefore, single pixel may include 16*16 distinct ratios of red (R), green (G), and blue (B), resulting in a total of more than a million sets and more than a million colors.

Photographs may develop too huge to deliver by Merit Internet connection when working with larger, higher-bit-depth photos. Techniques should be used to reduce the file size of the image so that it can be displayed in an acceptable amount of time.

These methods reduce document sizes by using numerical formulae to assess and condense image data. Compression is the name given to this procedure. Lossy and lossless compression is the two methods used in photographs. Both techniques conserve storage space, although they use different processing techniques.

Lossy compression reduces the size of the final document by throwing away extra photograph information from the source photograph. It can remove elements that too compact for a person's eye to perceive, producing approximate matches to the original image but not an exact match.

No data from the original image is ever lost during lossless compression. Compression is important when selecting a steganographic algorithm. Lossy compression techniques decrease the size of image documents, but they also raise probability of installed information perhaps invisible because more photography information will be eliminated.

Lossless compression does not limit the picture to such a small document size, but it does preserve the actual digital picture intact without the risk of loss.

9.8.1 Physical security in information security

Physical security is to protect people, property, physical assets, software, network connections, information against physical situations, natural disasters, theft, flood and fire in company, government organisation, or other entity [14].

Physical security consists of a number of components, including the following:

- It might be necessary to implement alternate physical security measures. When necessary, take into account if window bars, anti-theft cabling, locks, anti-virus and malware protection, strong password, encryption and motion detectors are required.
- It supports adequate personnel training to maintain fire extinguisher or install smoke alarms in the case of fire emergency.
- Sustain adequate room temperature and humidity level.

- It is possible to reduce the amount of unnecessary items that could endanger a secure space.
- Sensitive waste can be properly and carefully disposed of to maintain confidentiality.
- Confidential data can be properly labelled and provided with the necessary security procedures from common carriers when sending or receiving confidential data.
- It can be utilised to keep essential systems apart from non-essential systems.
- Use firewall for Computer accessories.
- It can be used to secure cabling, plugs, and multiple cables from foot traffic.
- In the case of emergency plan for an alternate place for natural disasters.
- To maintain the voltage use surge protectors.
- Computer equipment needs proper yearly upkeep and repairs.
- Appropriate procedures for backing up system data and applications.
- Establish system backup method and schedule.

9.9 DATABASE SECURITY IN INFORMATION SECURITY

It denotes all procedures taken to protect a databank against illegal access as well as damaging cyber threats and attacks. A level of information security is database security. It primarily addresses concerns related to data remanence, storage encryption, and physical information protection [15].

It is a blanket word that covers a wide range of practices, instruments, and approaches that offer security in a database context. A database management and other information security expert typically plan, implements, and maintain database security.

The World Wide Web offers a reliable, affordable, easily accessible, and quick method of data distribution. Although it makes distribution relatively easy, it is crucial to ensure that only users who have access permissions to the data should be able to access it.

Corporate information security has become crucial as a result of certain firms employing dynamic web sites that rely on databases. Prior to now, viewing the data required stringent database access or specialist client software, but now a basic web browser is sufficient to read data in a database that is not properly protected. Information security is thus in a vulnerable state. As a result, there is a greater risk of security dispersion the more a computer corporation switches from client-side processors to the internet.

Due to the nature of Intranet/Internet information access, security database experts must rely on network management to construct firewalls or other structures to protect local information. Nevertheless, the database administrator (DBA) must perform some security services. This lesson will examine the fundamental security concerns that lie within the purview of the DBA, who must then come up with database-savvy solutions.

Some security flaws are inaccessible because attackers must wait for a remedy and manufacturers do not want the negative publicity. The question of whether exposing security flaws in the open encourages or aids in preventing such attacks is still up for debate.

The most secure database that can be imagined should be kept in the tightest-latched bank or nuclear-proof bunker, installed on a solitary computer without an Internet or web connection, and guarded around-the-clock.

A database server must be realistic about potential hazards if it is to continue providing services, which typically include security issues. It ought to plan for failure and avoid gathering highly sensitive data in a database that is easily breached or accessed by criminals.

9.10 NETWORK SECURITY IN INFORMATION SECURITY

It is the study of controlling firewalls and rules to safeguard the network. In order to prevent unauthorised users from gaining access, network administrators employ a number of policies and procedures under the umbrella of network security management.

Access is restricted by a number of security-related regulations. The procedure establishes a secure network, ensures its protection, and controls network services. The fundamental type of network security enables users to grant access to people by providing them with a password or ID.

Users who have access to network data are under the control and authority of network administrators. These users can access data and applications under their control by using a password or login ID. Depending on the needs of the network, networks can be configured as either private or public. Public and private networks that are utilised for a variety of purposes, including at the job site for communications and business transactions, fall under the broad umbrella of network security.

Authentication, which often involves a username and password, is the first step in the procedure. One-factor authentication is typically referred to be a quick authentication procedure that includes authenticating a login or password. A security device, token, card, or phone are examples of items that users typically keep and include them in two-factor authentication processes. The three-factor authentication, the ultimate classification, incorporates a method like a fingerprint or retina scan.

The need for network security management should be ascertained by a network administrator. Small businesses just need a basic security system; however large organisations or enterprises may need additional security resources to thwart assaults or unauthorised users.

Small enterprises involve common firewall and simple light-weight processes administration solution. Small business networks use anti-virus software or an anti-spyware package. Management has strong password and premier level of security. Since most home users don't use this feature, small company owners should think about changing the network's default SSID name and turning off the SSID broadcast feature.

A strong proxy and firewall should be part of a government network security system to prevent illegal access from both inside and outside. Strong encryption and anti-virus software should be part of this network infrastructure.

Administrators need to make sure that hardware is kept in secure locations. To keep their private network hidden from other users, users of government networks should set up a private network that is only accessible by network hosts. Web servers that are situated in a DMZ and within a secure wireless range should also be a part of government systems.

9.11 NETWORK TYPES

In networks can be classified into three categories:

- **LAN** – Local area network connects computers in a restricted area, like office, business, school, or other organisation. Because of this, it only exists in that particular place, such a home network, business network, school network, etc.
- It may consist of a wired, wireless, or hybrid network. Ethernet cable offers boundary to connect numerous devices such as routers, bridges, switches, and computers, is typically used to connect the devices in a LAN.
- For instance, it can create a LAN at the home, workplace, etc. utilising a single router, a few Ethernet connections, and computers.
- **Metropolitan area network (MAN)** – A metro area, a town, and a wide geographic area are all covered by the high-speed MAN network. In order to install it, local telephone exchange lines and routers are used to connect the local area networks. It may be run by a private firm or offered as a service by a business like the neighbourhood phone provider.
- **WAN** – Wide area network typically set up using phone lines, fibre optics, or satellite connections and is not restricted to an office, school, city, or municipality. Large businesses, such as banks and multinational corporations, frequently utilise it to connect with their users and branches around the sphere.
- **Value-added network (VAN)** – A value-added network (VAN) is a private network operator that is contracted by an organisation to support various network services or to enable electronic data interchange (EDI). Public networks known as VANs bring value by facilitating access to commercial information and software as well as data transmission.

- **Internet** – A global network of computer networks is known as the Internet. It is a global interconnectedness of both large and small networks. It is, in other words, a vast network of networks that connects millions of computers.
- **Intranet** – A network using common IP, TCP/IP & HTTP can be used to connect an affiliated set of clients to form an intranet.
- **Extranet** – The term "extranet" can be used to refer to an intranet that allows regulated access by authorised parties and allows for restricted, well-ordered, and protected communication between work place intranet and specified, legal consumers in remote locations. The intranet and extranet are similar to LAN and WAN on the web.

9.12 TYPES OF AUTHENTICATION IN INFORMATION SECURITY

- Authentication is the process that enables a data sender and recipient to vouch for one another. There is no trust in the activities or data supported by either side if the sender and recipient of the data cannot adequately authenticate one another.
- Authentication can be really easy or involve a challenging and secure approach. Transmission of common PIN between entities seeking to validate one another is the simplest method of authentication.
- Authentication describes granting the network's resources access to just those users who are allowed to do so. It offers a mechanism where the access control structure uses some method to verify the claimed identifier.
- Individual verification is to guarantee only authorised users are accessing the services provided. There are several methods of authentication, including the ones listed below:
- **Admission resistor** – Structures and regulations made to limit access to computer resources to authorised persons only.
- **Documentation** – Strategy in which a reserve asserts (recognised) distinct and singular identify.
- **Authorisation** – It can establish the rights connected to the verified identity.
- **Security** – A system's capacity to protect information, facilities, and assets from exploitation by unauthorised users.
- **Privacy** – A system's capacity to protect the locations and identities of its users from unwanted acknowledgment.
- **Smart card** – Size is like a PAN card that delivers when associated to system and make payments. Hardware tokens are frequently employed in transaction especially Internet-based.
- **E-voting** – Electronic voting is another name for e-voting. It is an electronic system for both casting ballots and tallying results. Votes and

ballots may be transmitted over the Internet or a private computer network connected to a telephone.

- **Biometric authentication** – Alternatively, leave genuine user authentication. It deals with identifying a person based on behavioural or physical characteristics. This is very beneficial for illiterates.
- Physiological biometrics employ face, ear, finger-print, voice, finger geometry, palms, hand veins, hand geometry, iris connection, and now only brain waves are used. Behaviours' biometrics include gait, speech, signature, and keystroke search.
- **Data integrity** – In order to prevent unauthorised parties from changing information, data integrity refers to the consistency and accuracy of the information. Therefore, the data that is received should be identical to the data that was provided.

Protecting data transfer procedures is required to stop any alterations to these data, whether they are deliberate or accidental. The viability of these data or information will be impacted by any damage or drop in information. It stops being useful and unsafe to use. Data can be processed in a number of ways, including encryption.

9.13 TYPES OF FIREWALLS IN INFORMATION SECURITY

There are various types of firewalls, which are as follows:

9.13.1 Traditional network firewalls

By preventing undesirable traffic from entering the corporate network, packet-filtering network firewalls enable crucial network protection. They operate by employing a set of security rules for network firewalls to determine whether to allow or restrict access to the network.

It entails limiting access to some traffic to traffic going to specific ports that correspond to software running on the business network and allowing or disallowing access to data coming from certain IP addresses or utilizing certain protocols.

9.13.2 Circuit-level gateways

Another simplified sort of firewall is a "circuit-level gateway," which can be easily set to allow or deny traffic without using a lot of computational power. These firewalls typically examine TCP (transmission control protocol) connections and sessions at the session-level of the OSI model. Circuit-level gates are built to guarantee the security of the established sessions.

The majority of the time, anticipated firewalls or security applications employ circuit-level firewalls. These firewalls inspect transaction-related data but do not test for genuine data, similar to packet-filtering firewalls. As a result, data that contains malware but uses the correct TCP connection will pass through the gateway. Circuit-level gateways are not treated as safe to secure our systems because of this.

9.13.3 Proxy service firewalls

By filtering communications at the application layer, the proxy service firewall is a device that enhances network security. In essence, it acts as a gateway or middleman between the internal network and external network servers. It also goes by the name of a gateway firewall. It is safer since it inspects incoming traffic using stateful and deep packet inspection techniques.

9.13.4 Unified threat management (UTM) firewalls

A programme known as a unified threat management firewall links the capabilities of an SMLI firewall with those of antivirus and intrusion avoidance. The UTM services umbrella might include additional services like cloud management.

9.13.5 Next-generation firewalls (NGFWs)

In comparison to packet-filtering and stateful inspection firewalls, next-generation firewalls are more sophisticated. They have higher security levels and go beyond simple packet filtering to fully analyse a packet. It entails looking at a packet's contents and source in addition to its header. Advanced malware can be blocked by NGFW, which can handle more sophisticated and underdeveloped security threats.

9.13.6 Network address translation (NAT) firewalls

Firewalls that use network address translation, often known as NAT, are primarily made to access Internet traffic and deny access to some undesirable connections. These kinds of firewalls typically conceal the devices' IP addresses, making them safe from attackers.

9.13.7 Cloud firewalls

Cloud firewalls, also known as FAAS, are generated whenever a cloud-based technology is used to establish a firewall (firewall-as-service). Cloud firewalls are often managed and operated online by outside suppliers. This kind of firewall is seen as being equivalent to a proxy firewall.

9.14 OPENSTEGO TOOL FOR STEGANOGRAPHY

Steganography is covert means of sending communications that are undetectable. It comes in a variety of forms. Any one of these forms can be used to conceal data. Let's use the OpenStego tool for image steganography and the AES method for cryptography in this chapter.

OpenStego is the tool for steganography, which is used to conceal the original data with the help of image. This tool provides two functionalities, which are data hiding and watermarking. We discuss how the data hiding functionality works well.

9.14.1 A proposed technique for combination

In the proposed technique, the secret code is covered by the image. After the sender sends the covered file and key exchange, the receiver receives the covered file and key. At last receivers convert the secret message into original message.

9.14.2 Crypto module

The following steps are considered for encrypting the data in crypto module as shown in Figure 9.1:

*   Encrypt the actual data.
*   ApplyAESalgorithmusing128bitkey.
*   Generate Cipher Text.

9.14.3 Crypto module (reverse process)

The following steps are considered for retrieving the original text in crypto module:

*   The first step is going to get the above received cipher text.
*   The next level is the reverse process using the AES algorithm.
*   Finally, get the original message.

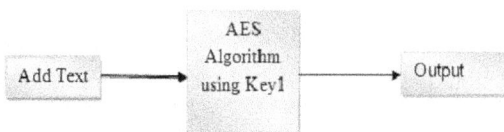

Figure 9.1   Crypto module.

9.14.4 Stego module

Stego module has the following steps, which were used for hiding the above generated Cipher text.

- Take some sample alphabets as an input.
- Jumble the alphabets using security key.
- Acquire grey scale image.
- Discover the LSB of the image.
- Conceal the cipher by altering LSB.
- Discover the Stego Image.

9.14.5 Stego module (reverse process)

The following steps are used to retrieving the cipher text in Stego module:

- Take LSB of the Stego image.
- Retrieve bits of the hidden alphabets from LSB.
- Construct the seven alphabets using the key.
- Retrieve the original alphabets.

9.14.6 Implementation details

This project is mainly developed in Python. Here mainly two modules are involved.

a. Crypto module – in crypto modules the AES algorithm is used for cryptography.
b. Stego module – In the Stego module the LSB algorithm implementation module is used for steganography.

9.14.7 Algorithm for the proposed system

In the proposed plan all the original messages are encrypted with the secret key. After that the secret message is to hide within the image using the LSB algorithm.

   The following steps for the algorithm are described below and in Figure 9.2.

9.14.8 How secure is the proposed system?

The proposed system is to provide highly secured communication. Here we used the combination of two highly secured techniques.

a. AES for cryptography.
b. LSB for steganography.

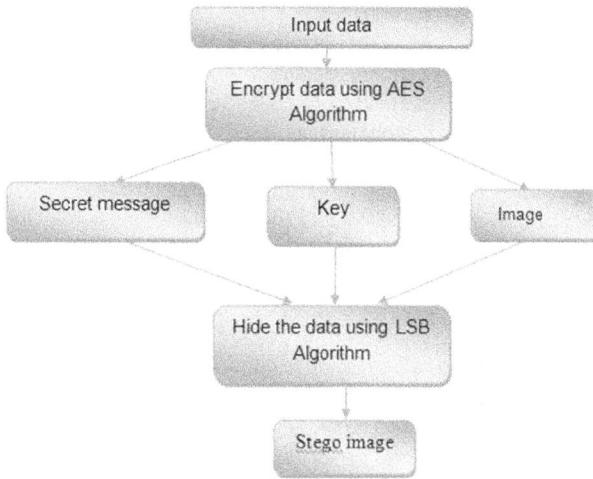

Figure 9.2 Proposed system for hiding text.

9.15 CONCLUSION

In this chapter, the first technique AES algorithm for cryptography is used to convert the original data into secret code. The second steganography technique uses the LSB algorithm to hide the confidential code into a picture. It also includes the Stego tool, which is also used for processing above cryptography with combined steganography algorithms for secured communications. This is highly secured for our growing Internet communications.

REFERENCES

1. Seth Dhawal, L. Ramanathan, Abhishek Pandey, Security Enhancement: Combining Cryptography and Steganography, *International Journal of Computer Applications*, vol. 9, no. 11, 0975–8887, November 2010.
2. Ramanan, M, L. Singh, A. Suresh Kumar, A. Suresh, A. Sampathkumar, V. Jain, N. Bacanin, Secure Blockchain Enabled Cyber–Physical Health Systems Using Ensemble Convolution Neural Network Classification. *Computers and Electrical Engineering*, vol. 101, 108058, 2022.
3. Aung Pye Pye, Tun Min Naing, A Novel Secure Combination Technique of Steganography and Cryptography, *International Journal of Information Technology, Modeling and Computing (IJITMC)*, vol. 2, no. 1, February 2014.
4. shaSethia Pratik, V. Kapoorb, A Proposed Novel Architecture for Information Hiding in Image Steganography by using Genetic Algorithm and Cryptography, *International Conference on Computational Science*, vol. 87, 61–66, 2016.

5. Suresh Kumar, A, S. Jerald Nirmal Kumar, Subhash Chandra Gupta, Anurag Shrivastava, Keshav Kumar, Rituraj Jain, et al., "IoT Communication for Grid-Tie Matrix Converter with Power Factor Control Using the Adaptive Fuzzy Sliding (AFS) Method", *Scientific Programming*, vol. 3, 2022.

6. *Proposed System for Data Hiding Using Cryptography and Steganography*, *DiptiKapoorSarmah, NehaBajpai.

7. AksharaSree Challa, Dr. B. Indira Reddy, Role of Cryptography and Steganography In Securing Digital Information: A Review, © 2018, *IJCRT*, vol. 6, no. 2, April 2018. ISSN: 2320-2882

8. Phapale Anuja, Rijil Daniel, Pranav Deshmukh, Dhanesh Lunkad, YogeshThadani, Image Based Steganography Using Cryptography, *International Journal of Technology Engineering Arts Mathematics Science*, vol. 1, no. 1, 24–27, January–June 2021.

9. Saji, Akhil K, Ansa Alex, Remy Raju, Rincy Roy Oommen, Combining Cryptography and Steganography for Data Hiding in VideosAnjuLukose, *IJIRT*, vol. 5, no. 12, May 2019. | ISSN: 2349-6002

10. Naveen Kumar H. N, A. Suresh Kumar, M. S. Guru Prasad, Mohd Asif Shah, "Automatic Facial b Expression Recognition Combining Texture and Shape Features from Prominent Facial Regions", *IET Image Processing*, https://doi.org/10.1049/ipr2.12700, 2022

11. Neetha Francis, Information Security using Cryptography and Steganography, *International Journal of Engineering Research & Technology (IJERT)*, vol. 3, no. 28, 1–6.

12. MallikharjunaRao, G. Information Security using Cryptography and Image Steganography, Inter*national Journal of Recent Technology and Engineering (IJRTE)*, 9(2), July 2020. ISSN: 2277-3878 (Online)

13. Jan Aiman, Shabir A. Parah MuzamilHussan, Bilal A. Malik, Double Layer Security Using Crypto-Stego Techniques: A Comprehensive Review, *Health and Technology*, vol. 12, 9–31, 2022.

14. Varsha, Dr. Rajender Singh Chhillar, Data Hiding Using Steganography and Cryptography, *International Journal of Computer Science and Mobile Computing*, ISSN 2320–088X, ISSN 2320–088X

15. Marwa, E. Saleh, Abdelmgeid A. Aly, Fatma A. Omara, Data Security Using Cryptography and Steganography Techniques, *(IJACSA) International Journal of Advanced Computer Science and Applications*, vol. 7, no. 6, 1–8.

Chapter 10

Ubiquitous and transparent security

Intelligent agent framework for secure patient–doctor modelling systems

Manivel Kandasamy[1], Raju Shanmugam[1],
Palakshi Sinha[1], Tanya Chhabhadiya[1],
and A. Suresh Kumar[2]
[1]Unitedworld Institute of Technology, Karnavati University, Gandhinagar,
Gujarat, India
[2]School of Computer Science and Engineering, Jain (Deemed to be University),
Bengaluru, Karnataka, India

10.1 INTRODUCTION

E-health has advanced significantly thanks to the incredible combination of the Internet of Things (IoT) with traditional health monitoring technologies [1]. One noteworthy achievement is patient–doctor modelling systems, which use sophisticated technologies to enable seamless interactions between healthcare practitioners and patients. Patients are now constantly being watched over thanks to Internet of Things (IoT) gadgets. Despite their modest size, these gadgets are capable of monitoring the patients who are wearing them. The patient can be remotely observed during an emergency, and appropriate actions can also be conducted. These devices can concurrently identify irregularities and provide emergency assistance because they are constantly connected to the Internet [1]. By streamlining medical procedures and enhancing patient interaction, these modelling systems hope to raise the standard of care. To maintain the privacy, integrity, and accessibility of sensitive patient data, these developments are accompanied by significant security problems that must be resolved.

Medical records, diagnoses, treatments, and other sensitive information are all handled in significant quantities in the healthcare industry. The potential for unauthorised access, data breaches, and malicious actions increases as patient–doctor modelling systems become more integrated and available across a range of platforms and devices. The sophistication of cyberattacks against healthcare organisations has increased, and the possible repercussions, including harmed patient privacy and disrupted healthcare services, are worrisome.

There is an increasing need for robust and intelligent security frameworks that offer all-encompassing and transparent protection for patient–doctor

DOI: 10.1201/9781003362685-10

modelling systems in order to reduce these security threats. In order to build a comprehensive and reliable security posture, ubiquitous security involves the integration of security measures throughout the entire system architecture as opposed to isolated components. An effective information interchange between patients and doctors is made possible by transparent security, which refers to a security framework that does not obstruct the system's usability and transparency

10.1.1 Ubiquitous computing: Transforming the future with connected technologies

The idea of ubiquitous computing has grown significantly in importance in today's rapidly evolving technological environment. Ubiquitous computing, also referred to as pervasive computing or the Internet of Things (IoT), is the seamless incorporation of computing devices and technologies into our daily lives. It imagines a time when computing permeates our surroundings, fostering a networked environment that boosts convenience, productivity, and efficiency. In this chapter, we will examine the fundamental ideas and technological underpinnings of ubiquitous computing and consider how it might affect different facets of our daily lives.

At its core, ubiquitous computing seeks to establish a setting in which computing resources are seamlessly woven into our daily routines. It imagines a world in which information and communication technologies are integrated into a variety of commonplace objects and environments rather than just traditional computing devices like desktop computers and smartphones. The proliferation of interconnected devices and sensors creates an ecosystem where computing is pervasive, from home appliances and wearables to cars, buildings, and public spaces.

Applications and impact: Ubiquitous computing has the power to transform a number of industries, including the following:

1. Smart homes: With ubiquitous computing, homeowners may remotely manage and keep an eye on a variety of aspects of their homes. Smart homes use interconnected devices and sensors to provide a comfortable and secure living environment, from regulating temperature and lighting to managing security systems.
2. Healthcare: In the field of healthcare, ubiquitous computing can enhance remote healthcare services, prescription adherence, and patient monitoring. Vital sign monitoring is possible with wearable technology and real-time data sharing between patients and healthcare professionals is made possible through linked health platforms, resulting in proactive and individualised healthcare.
3. Transportation: Smart traffic management, networked cars, and autonomous driving are all made possible by ubiquitous computing, which is revolutionising the industry. Real-time traffic monitoring, effective

routing, and improved safety features are made possible by sensors implanted in infrastructure and cars.

4. Commerce and retail: Personalised shopping experiences, inventory management, and targeted marketing are made possible by ubiquitous computing, which is transforming the retail sector. Retailers can give location-based promotions and offer seamless checkout experiences thanks to the beacon and RFID technologies.

5. Smart Cities: The foundation of innovative city efforts is ubiquitous computing, which boosts the effectiveness of public services, energy management, and environmental sustainability. Examples of how ubiquitous computing might alter urban life include intelligent street lighting, waste management systems, and optimised transportation networks.

10.1.2 Healthcare and ubiquitous computing: Changing the future of patient care

IoMT, or the Internet of Medical Things, is the result of IoT use in the healthcare industry. Offering services like real-time health monitoring, patient information management, tracking illness and epidemic outbreaks, diagnostic and treatment support, digital medication, etc., makes life easier for both patients and doctors. Patients' various physiological characteristics can be continuously monitored by wearable IoT devices or customised biosensors [2]. The healthcare sector is undergoing a change thanks to ubiquitous computing. Ubiquitous computing provides innovative solutions that increase patient care, boost medical monitoring, and optimise healthcare services by integrating computing technologies into the healthcare ecosystem. This brief will examine the main uses and advantages of ubiquitous computing in healthcare.

10.1.2.1 Remote patient monitoring

Remote patient monitoring is made possible by ubiquitous computing, giving medical professionals access to real-time information regarding a patient's vital signs, medication compliance, and general health state. Sensor-equipped wearables gather information such as heart rate, blood pressure, temperature, and activity levels and transmit it to medical specialists. Particularly for patients with chronic diseases, this technology offers early detection of irregularities, facilitates prompt therapies, and decreases the need for repeated hospital visits.

10.1.2.2 Enhanced medication management

By ensuring that patients take their prescribed drugs correctly and on schedule, ubiquitous computing technologies help to improve medication management. Sensors and connections are used by smart pill dispensers and medication reminder devices to notify patients when it is time to take their meds. These systems can also track medication adherence and offer dose guidelines, enabling healthcare professionals to take appropriate action. Ubiquitous computing improves patient safety and overall treatment results by lowering drug mistakes and adherence issues.

10.1.2.3 Telemedicine and virtual healthcare

Ubiquitous computing makes telemedicine and virtual healthcare possible by allowing for distant consultations and monitoring. Healthcare practitioners can remotely evaluate patients, offer medical advice, and track their progress using video conferencing, secure messaging, and data-sharing platforms. Access to healthcare services is increased because of this technology, especially for people who live in rural locations or have limited mobility. Telehealth applications including remote patient counselling, mental health support, and rehabilitation services are also supported by ubiquitous computing.

10.1.2.4 Ambient assisted living

Ubiquitous computing encourages ambient assisted living, allowing elderly or disabled people to live independently while receiving the care and support they need. Sensors, voice assistants, and wearable technology in smart home systems allow for the monitoring of everyday activities, the detection of falls, and the provision of prescription, appointment, and routine reminders. In the event of an emergency, these systems can also notify carers or emergency services, protecting the safety and well-being of anyone in need of assistance.

10.1.2.5 Intelligent health monitoring systems

Intelligent health monitoring systems are made possible by ubiquitous computing technologies like sensor networks and data analytics. These systems gather and examine massive amounts of patient data in order to find patterns, recognise trends, and give healthcare personnel useful information. Early warning indications of deteriorating health issues can be identified using machine learning algorithms, allowing for preventive therapies. Through the use of real-time data and sophisticated analytics, ubiquitous computing improves patient outcomes, treatment efficiency, and diagnosis accuracy.

10.1.3 Future scope and applications of intelligent agents in healthcare

The healthcare sector has rapidly evolved in recent years as a result of technological breakthroughs. The use of intelligent agents, which are software entities capable of autonomous decision-making and problem-solving, is one field with a lot of potential. These intelligent agents have the potential to fundamentally alter healthcare delivery while also benefiting patient outcomes and general effectiveness. Let's explore the potential uses of intelligent agents in healthcare.

1. Personalised medical attention: Massive volumes of patient data, like medical records, genomic information, lifestyle factors, and environmental factors, can be collected and analysed by intelligent agents. These agents can produce insightful recommendations for personalised care plans that are catered to the unique requirements of each person by using machine learning algorithms. This enables healthcare professionals to offer more individualised and efficient therapies, improving patient outcomes.

2. Clinical decision support: Intelligent agents can act as trustworthy clinical decision support systems, helping medical practitioners make informed, research-based judgments. These agents are able to deliver real-time alarms, detect potential hazards, provide treatment alternatives, and even forecast adverse outcomes by connecting with electronic health records and continuously monitoring patient data. This increases the knowledge of healthcare professionals, lowering errors and improving the standard of service.

3. Remote patient monitoring: This is made possible by intelligent agents that gather and examine data from wearables, sensors, and IoT-enabled devices. These agents can recognise early warning indicators, quickly alert healthcare personnel, and start necessary interventions by continually monitoring vital signs, medication adherence, and other health data. In addition to reducing readmissions to hospitals and increasing patient participation and satisfaction: This promotes proactive and preventive care.

4. Health education and engagement: Intelligent agents can serve as virtual health coaches who give patients individualised health advice. They can provide advice on managing medications, altering one's lifestyle, and using illness-prevention techniques. Additionally, by answering questions, providing assistance, and encouraging healthy behaviours through conversational interfaces, these agents can interact with patients. As a result, patients are given the ability to actively take part in their own care and make knowledgeable decisions regarding their health.

5. Healthcare resource optimisation: Allocating and utilising healthcare resources can be made more effective by intelligent agents. These agents can locate problems, estimate demand, and offer suggestions for resource optimisation by analysing data on patient flow, bed occupancy, staff availability, and medical equipment usage. By doing this,

healthcare organisations can increase operational effectiveness, cut costs, and boost patient satisfaction and accessibility.

10.2 INTELLIGENT AGENT-BASED SECURITY MECHANISMS

There are many advantages to the growing digitisation of healthcare data, but it has also sparked worries about patient privacy and data security. To protect the confidentiality between doctors and patients from various threats like cyberattacks, data breaches, medical identity theft, and third-party risks the need for intelligent agent-based security mechanisms arises. Various mechanisms have emerged as a possible response to the expanding cyber threats aimed at health data [3]. This strategy equips healthcare systems to proactively identify, stop, and react to security breaches in real time by utilising autonomous software entities known as intelligent agents. Due to their distinct benefits and capacities in addressing the complex and dynamic nature of cybersecurity concerns in the healthcare domain, intelligent agent-based models are being utilised more frequently for securing healthcare data.

In general intelligent agent-based security mechanisms use autonomous software entities known as "intelligent agents" to improve a system's security. In order to adapt and learn from new information, these intelligent agents frequently use machine learning techniques. They are capable of monitoring, analysing, and making decisions based on data and expertise. In order to spot anomalies or potential security breaches, these agents can function in real-time, continuously monitoring user behaviour, network traffic, and system activity. They can also examine prior data to draw lessons from accidents and anticipate hazards in the future.

The common forms of attacks carried out thus far throughout the stage of data collecting include jamming-based attacks, collision-based attacks, de-synchronisation-based attacks, spoofing attacks, and selective forwarding attacks. Using such intelligent systems at an early stage of data collection can prevent such threats in the future. Characteristics of an intelligent agent-based security mechanism include the following:

1. Adaptability: Adapt to new threat patterns and environmental changes.
2. Collaboration: Work together and exchange information, providing a group defense against complex attacks.
3. Learning capabilities: These systems evaluate data using machine learning algorithms and draw lessons from the past, allowing them to get better over time.
4. Autonomy: Intelligent agents may make decisions on their own without human input, enabling prompt reactions to security problems.
5. Real-time monitoring: These controls keep an eye on the system constantly, sending out notifications and taking action in reaction to security incidents.

6. The defense that is proactive: Intelligent agents can foresee potential dangers and take precautions to safeguard the system before an assault happens.
7. Context awareness: Intelligent agents take into account context when making security decisions, such as user conduct, location, and time.

Several approaches can function as intelligent agent-based control systems for healthcare data security. Such systems use intelligent agents to monitor user activity, improve access control, and find security issues. Many security mechanisms are designed individually and then incorporated into the system for better functionality and result. Below are some security mechanisms:

1. Predictive analytics: Agents that have access to predictive analytics are able to foresee possible security risks and weaknesses. They are able to evaluate patterns, spot anomalies, and anticipate potential attack vectors, enabling the implementation of proactive security measures prior to incidents.
2. Anomaly detection: Machine learning techniques can be used by intelligent agents to identify anomalies in the access to and use of data. Agents can detect odd activity, such as strange data searches or access from unusual places, and flag them for further examination by continuously monitoring data exchanges.
3. Adaptive access control: Intelligent agents have the ability to adapt access privileges on the fly in response to user behaviour and circumstances. Agents can spot deviations and make instantaneous access control choices to stop unwanted access by learning from previous access patterns and user behaviours.
4. Self-healing systems: By using intelligent agents, healthcare data systems can react to security incidents on their own. For instance, if a hacked user account is found, the agent can start isolating the privileges of the infected user and carry out the required cleanup procedures.
5. Threat hunting: Agents might function as preemptive "threat hunters" that scour the healthcare data system for any potential security issues. To stop security breaches, they can search for signs of compromise, well-known malware signatures, or other unusual activity.
6. User behaviour profiling: Agents can develop user behaviour profiles to look for account compromise or identity theft. Agents can spot behavioural anomalies and generate alerts for more inquiry by examining user interactions.
7. Context-based authorisation: Agents can make better authorisation decisions by using contextual data, such as location, time, and the access device. This makes sure that only reliable and acceptable sources are used to grant access.
8. Automated patching and updates: The method of implementing security updates and patches to the healthcare data system can be automated

by intelligent agents. As a result, the risk of exposure is decreased by ensuring that known vulnerabilities are swiftly fixed.

9. Secure communication protocols: To safeguard data transferred between various elements of the healthcare data system, agents might adopt secure communication protocols. This stops data from being intercepted and altered while in transit.

10. Data encryption and tokenisation: Intelligent agents can control encryption keys and tokenisation processes to secure healthcare data. The risk of data disclosure in the event of unwanted access is decreased by agents by encrypting vital data and substituting it with tokens.

Intelligent agent models have become a powerful and flexible method for protecting sensitive data in the field of data security. These models may dynamically monitor, assess, and react to security risks in real-time by using autonomous software entities called intelligent agents. There are several models available like Intelligent Behavioural Access Control (IBAC), Attribute-Based Encryption, Context-Aware Role-Based Access Control (CARBAC), and Intelligent Agent-Based Anomaly Detection (IABAD), Intelligent Agent-Based Intrusion Detection System (IIDS), etc.

10.3 INTELLIGENT BEHAVIOURAL ACCESS CONTROL (IBAC)

IBAC is a data security technique that uses intelligent agents to continuously evaluate user activity and implement access restrictions based on risk evaluation. IBAC provides or denies access while utilising machine learning and data context to improve data security. Data gathering, behavioural analysis, intelligent agents, policy enforcement, and access control rules are all part of the architecture. IBAC has uses in corporate networks, banking, and healthcare. For an implementation to be successful, however, issues including incorrect grant or deny access due to anomalies, shifting user behaviour, privacy issues, and complexity must be taken into account. Despite its drawbacks, IBAC provides a proactive strategy for guarding against growing cyber risks and maintaining confident access control [4].

10.4 INTELLIGENT MULTI-AGENT SYSTEM FOR HEALTHCARE DATA SECURITY (IMASH)

A cooperative security architecture called Intelligent Multi-Agent System for Healthcare Data Security (IMASH) makes use of a network of intelligent agents to protect healthcare data. To maintain safe access management, agents monitor data, look for anomalies, and exchange threat intelligence encryption and machine learning is used. Patient records, medical IoT devices, and health information transfers can all be protected with IMASH.

Despite scaling issues, communication costs, and installation complexity, IMASH provides a collective defense strategy that is aware of its environment. IMASH improves healthcare data security by utilising the knowledge of various agents, facilitating proactive threat detection, and guarding delicate patient information against rising cyber threats.

10.4.1 Attribute-based encryption

A cryptographic method called attribute-based encryption (ABE) provides secure data transfer based on predetermined attributes or criteria. ABE allows for the granting of access to encrypted data based on matching attributes, such as user roles, attributes, or predetermined policies. In circumstances where access is controlled by intricate attribute-based criteria, it enables fine-grained access control and adaptable data sharing, encrypted cloud storage, data sharing in healthcare systems, and encrypted communication are all areas where ABE finds use. Key management, scalability, and significant performance overhead are issues with ABE, despite the fact that it provides improved access control and data protection. However, ABE offers a strong encryption method based on attribute-based regulations for secure and controlled data sharing [5].

10.4.2 Transparent security in patient–doctor modelling systems

Modern technologies have made it possible to put innovative solutions into practice to improve the standard of living for people. In order to learn more and address health-related issues, researchers looking at the development of technology have discovered and assessed sources of health information. At every level of the healthcare system, the development of integrated healthcare technology, therefore, has the potential to increase productivity and improve patient outcomes. Through reliable patient safety controls, ubiquitous data access, remote inpatient monitoring, quick clinical interventions, and decentralised electronic health records, the development of new electronic health (e-Health) application systems can address some issues related to traditional healthcare systems [6].

The following diagram (Figure 10.1) depicts the basic system architecture for healthcare integrated with ubiquitous computing

The earlier-described diagram can be modified to emphasise the significance of clarity and transparency in the healthcare process in the context of transparency in a patient–doctor modelling system. Let's see how each phase incorporates transparency:

Step 1: Transparent medical data analytics, smart pill dispensers, and sensor-based wearables.

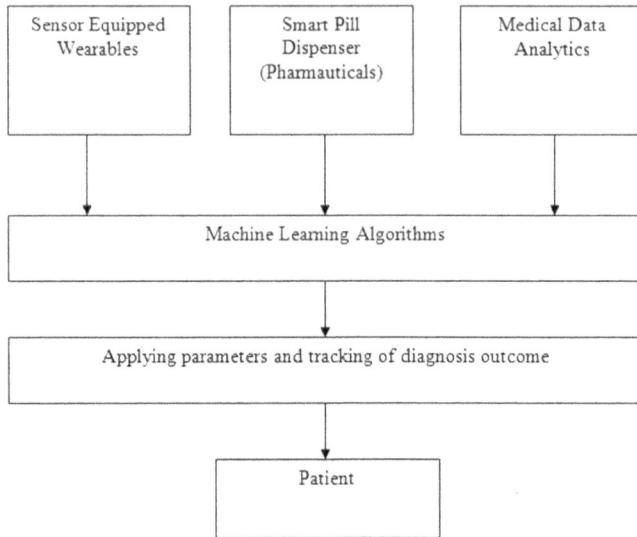

Figure 10.1 System architecture for healthcare integrated with ubiquitous computing.

Transparency starts with the process of gathering data from wearables with sensors and automated pill dispensers. Patients must be made aware of the sorts of data being gathered, the reason for the data collection, and the ways in which their data will be utilised to enhance patient outcomes. This makes sure that patients are fully aware of the data being gathered and how it fits into their treatment plans.

Mehta et al. [7] constructed a mobile voice health monitoring system that takes advantage of the accelerometer sensor in smartphones.

In the device under examination, a tiny accelerometer serves as a voice sensor, and the smartphone serves as the platform for data collection. The patient's neck is wrapped in the system. Although the frame-based voice parameters were employed in this system, monitoring can also be done with the raw accelerometer data [7]. A multi-lead ECG health monitoring system based on a smartphone was proposed by Gao et al. [8].

Additionally, transparency should be considered when performing medical data analytics. Patients have a right to be informed about the methods used to analyse their data and the kinds of insights that are being produced.

Step 2: Transparent machine learning algorithm.

When using machine learning algorithms in healthcare, transparency is essential. Patients should be informed that these algorithms are being trained and improved using their data. While addressing any concerns regarding data privacy and security, it's critical to convey the advantages of using machine learning, such as improved diagnosis accuracy and personalised treatment regimens.

The machine learning system's limits and the possibility of biases or errors in the algorithms should be discussed openly by healthcare providers. In order to preserve a sense of trust and understanding in the procedure, patients should be educated about how decisions are made based on the output of these models.

Step 3: Applying parameters and tracking diagnosis outcome with transparency.

Explaining to patients how the individualised treatment plans are created using the knowledge obtained from the machine learning models is a crucial part of being transparent when implementing parameters. Healthcare professionals should be open and honest about the thinking behind their decisions, and patients should be involved in the decision-making process. Transparency is important when monitoring the results of a diagnostic since it allows patients to be informed about the patient's health status and how well their treatment is working.

Step 4: Patient.

The entire patient experience depends on transparency. Patients should be made aware of the benefits and purposes of using data analytics, smart pill dispensers, and sensor-based wearables to monitor and manage their health. The patient and healthcare professional feel a sense of cooperation as a result of this openness.

Additionally, patients should have the highest privacy and security when it comes to how their data is handled. Building confidence in the patient–physician relationship requires providing clear explanations about data-sharing procedures, informed consent, and data anonymisation (where applicable).

10.5 SECURE COMMUNICATION AND DATA EXCHANGE BETWEEN PATIENT–DOCTOR MODELLING SYSTEMS

People are more in need of health management and medical services as society develops and as living standards rise. Utilising mobile communication technology, such as tablets and smartphones, mobile healthcare, sometimes referred to as mobile health, can offer medical services and information. The ability to access a variety of health-related information (including physical examination, healthcare, disease evaluation, medical treatment, and rehabilitation) either at home or on the go is provided as an efficient solution [9].

Health authorities give the possibility of patient misidentification a lot of thought in order to try to prevent dangerous outcomes like prescription errors, wrong surgical treatments, etc. Health authorities give the possibility of patient misidentification a lot of thought in order to try to prevent dangerous outcomes like prescription errors, wrong surgical treatments, etc. [4]. This chapter conveys the proposal of using Blockchain-Enabled Patient Health Record (PHR) Platform.

In addition to protecting data confidentiality and privacy, a user-friendly and decentralised patient health record (PHR) platform built on blockchain technology can give patients quick access to their medical records. For a smooth patient experience, our platform makes use of the transparency and data integrity capabilities of blockchain while designing a user-centric interface. The technology provides a straightforward interface that makes it easy for patients to view their medical details. The dashboard gives the patient a clear picture of their health information and makes it simple for them to move between areas.

A decentralised blockchain network can be used to store patient health records. The data for every patient will be divided up, distributed across several machines, and encrypted to ensure maximum security and prevent single points of failure. The platform includes biometric authentication, such as fingerprint or facial recognition, to increase security without reducing convenience. With the help of a quick and simple biometric scan, this function enables patients to securely access their medical records. The platform offers patients a mobile app so they may access their PHR while on the go. Patients may easily examine their medical history, view test results, and manage appointments from their smartphones because of the app's user-friendly design.

The platform easily connects to wearable health gadgets like smartwatches or activity trackers. These devices instantly sync patient-generated health data with their PHR, giving them a complete picture of their health journey to the doctor as well as the patient, and can update any concern regarding their health actively throughout the day. Patients can easily access and manage their health records while taking advantage of the security and transparency provided by blockchain technology by implementing a Patient Health Record (PHR) Platform that is blockchain-enabled. Patient involvement is improved by the user-friendly interface, biometric authentication, and mobile app accessibility, enabling people to take an active role in their healthcare as shown in Figure 10.2.

10.6 IMPACT OF INTELLIGENT AND TRANSPARENT SECURITY

Our lives have been significantly impacted by intelligent and smart security technology, especially in the areas of safety and security. These cutting-edge technologies make better, more adaptable security measures possible by utilising machine learning, artificial intelligence (AI), data analytics, and the Internet of Things (IoT).

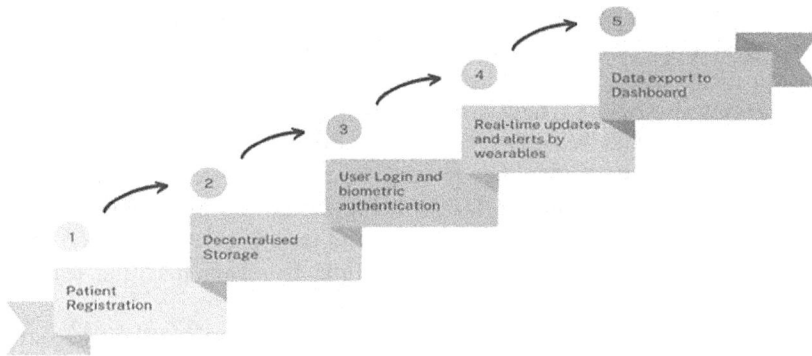

Figure 10.2   Flowchart for secure communication and data exchange.

Intelligent security systems are able to scan enormous volumes of data in real-time and spot trends and anomalies that could point to security concerns. This includes identifying odd behaviour in real-world settings (such as facial recognition or suspicious motions) as well as spotting potential cyber threats in virtual places. Compared to human operators, automated security systems can react to attacks far more quickly. To avoid or lessen security problems, for instance, AI-powered surveillance cameras can instantly warn security staff when they spot a potential breach.

To monitor and react to events, traditional security systems mainly rely on human operators. Unfortunately, mistakes made by people can result in serious security vulnerabilities. By automating tedious operations and making choices based on data-driven insights, intelligent security technologies reduce these mistakes. Smart security solutions are frequently highly flexible and adaptive to different situations in terms of scalability. These technologies can be modified to meet specific demands and expanded as conditions change, whether it's for a tiny office, a sizable industrial complex, or a smart city project.

Although intelligent security solutions may require a sizable initial investment, over time, costs may be reduced. Businesses and organisations can minimise their security costs by requiring less intensive human monitoring and offering more focused responses to problems. Access control, building automation, and energy management are a few examples of other smart systems that can be integrated with smart security technology. Because of this convergence, security can be approached more holistically and comprehensively and decisions can be made using information from a variety of sources. There are substantial privacy and ethical problems when it comes to smart security because it frequently includes gathering and analysing personal data. It continues to be difficult to strike a balance between maintaining safety and upholding each person's right to privacy.

Some security employees may lose their jobs as a result of the auto-mation of specific security functions by intelligent technology. But these innovations also open up new career paths for those in positions like system administrators, data scientists, and AI analysts. Smart security measures that are implemented openly may serve to prevent potential criminals or wrongdoers. Knowing that a place is being monitored by intelligence may deter illegal activity. The threat detection and response capabilities of smart security systems can be improved over time by using machine learning algorithms that can continuously learn from new data.

The learning process and internal perspective of an intelligent system toward the problem can differ from that of a human, generating dissonance, suggesting that system performance alone is not sufficient as a criterion. This is because overall performance indicators of ML models can be utilised to evaluate the recommendation effectiveness of an intelligent system [10].

Transformative security: Safeguarding healthcare data with intelligent behavioural Access Control.

It is crucial to ensure the confidentiality and privacy of sensitive health data while developing safe patient–doctor modelling systems. Passwords and biometrics are two common traditional access control strategies, but they have drawbacks and might be vulnerable to unauthorised access. Intelligent Behavioural Access Control (IBAC) is a cutting-edge idea that addresses these problems.

IBAC is a cutting-edge access control system that uses machine learning and artificial intelligence algorithms to identify and authenticate users based on their particular behavioural patterns and traits. IBAC continu-ously analyses dynamic behavioural aspects, such as typing habits, mouse movements, speech patterns, and even biometric data, to develop a thorough user profile rather than relying simply on static credentials like passwords or fingerprints. By utilising adaptive access control and intelligent profiling, it is possible to restrict access to sensitive patient data to only those who are authorised.

Below is a description of IBAC's function in secure patient–physician modelling systems.

1. IBAC analyses user behaviour during login attempts and compares it to the behavioural profile that has been previously stored. Access is allowed if the behavioural characteristics line up. Additional verifica-tion procedures may be activated in the event of suspicious behaviour, thwarting unauthorised access attempts.
2. Continuous monitoring: IBAC keeps track of user activity at all times during a session. Access can be immediately suspended or restricted

if behavioural patterns drastically differ from the established profile, reducing the risk of data breaches.

3. Transparent user experience: The IBAC authentication process is seamless and transparent for patients and healthcare professionals. Users are authenticated based on their typical behaviour, guaranteeing a user-friendly experience instead of obtrusive security procedures.

The illustration in Figure 10.3 shows a secure patient–doctor modelling system with an intelligent agent architecture deployed, with Intelligent Behavioural Access Control (IBAC) as its key component. The components of the diagram are as follows:

1. User authentication layer

   Patients and healthcare professionals are the system's two main users. It could use different methodologies for authenticating the user or the doctor.

2. Intelligent agent framework

   Based on the behavioural information gathered during login attempts and user interactions, these agents continuously learn about and update user profiles. The foundation for adaptive and secure access control is built on the behavioural data, which is used to construct and maintain personalised user profiles for each person.

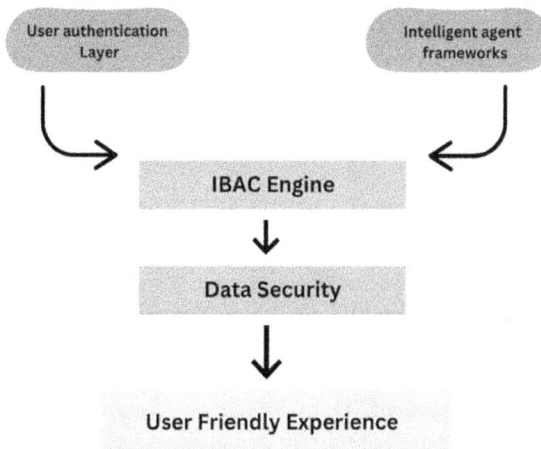

Figure 10.3   Basic architecture for IBAC.

3. IBAC engine

The IBAC engine evaluates and authenticates users based on their dynamic behavioural patterns by using artificial intelligence and machine learning algorithms. When a user attempts to log in and throughout their session, the IBAC engine continuously assesses their behaviour, providing access only if the behavioural characteristics match the defined user profile.

4. Data security layer

This layer denotes the safe handling and storage of patient data.

In order to protect patient data, the IBAC engine continuously observes user behaviour during the session and, in the event of suspect behaviour, briefly suspends or restricts access.

5. User-friendly experience

Here, users are made aware of the behavioural data collection and its application to user experience transparency, trust-building, and authentication.

Secure patient–doctor modelling systems have advanced significantly with the incorporation of Intelligent Behavioural Access Control (IBAC) into an intelligent agent architecture. When used for authentication, dynamic behavioural patterns improve security while delivering a transparent and user-friendly experience. The future of healthcare is primed for transformation by utilising linked technology and cutting-edge security mechanisms like IBAC – a future where patient data is protected and trust between patients and healthcare providers is nurtured.

10.7 CONCLUSION

The seamless connectivity and integrated systems made possible by ubiquitous computing are changing the healthcare industry. Intelligent agents are essential to this shift because they analyse patient data, help with diagnosis, and enhance clinical judgment. Intelligent agents have a potential future in healthcare, improving patient care and medical knowledge. Intelligent agent-based security methods are necessary to protect sensitive patient data and medical equipment. Data integrity is ensured and confidence is fostered through transparent security in patient–doctor modelling systems.

These steps have a deterring effect on potential wrongdoers and speed up danger detection while lowering errors. However, the use of technology in healthcare poses questions about ethics and privacy. It's critical to strike a balance between the potential for transformation and patient rights. Benefits can be enhanced while negative effects are minimised when privacy laws and ethical principles are followed.

In conclusion, the future of patient care is extremely hopeful as a result of the confluence of ubiquitous computing and healthcare. A safer, more effective, and integrated healthcare environment will result from appropriately integrating intelligent agents and transparent security. We can realise the entire potential of these advances and build a healthier society for everyone by putting a strong emphasis on privacy and ethics.

REFERENCES

1. Hussain, A., Ali, T., Althobiani, F., Draz, U., Irfan, M., Yasin, S., Shafiq, S., Safdar, Z., Glowacz, A., Nowakowski, G., Khan, M. S., & Alqhtani, S. M. (2021, March 18). Security framework for IoT-based real-time health applications. *Electronics; Multidisciplinary Digital Publishing Institute.* https://doi.org/10.3390/electronics10060719

2. Chattopadhyay, A.K., Nag, A., Ghosh, D., & Chanda, K. (2018, October 5). A secure framework for IoT-based healthcare system. *Advances in Intelligent Systems and Computing.* https://doi.org/10.1007/978-981-13-1544-2_31

3. Khan, Faizal & Reyad, Omar. (2019). Application of Intelligent multi Agent-based Systems for E-Healthcare Security. *Information Sciences Letters*, 8. 10.18576/isl/080204

4. World Health Organization. *Field Review of Patient Safety Solutions.* World Health Organization, Geneva, Switzerland, 2008.

5. P, P. K., P, S. K., & P.J.A., A. (2018). Attribute based encryption in cloud computing: A survey, gap analysis, and future directions. *Journal of Network and Computer Applications*, 108, 37–52. doi:10.1016/j.jnca.2018.02.009

6. Butpheng, C., Yeh, K. H., & Xiong, H. (2020, July 17). *Security and Privacy in IoT-Cloud-Based e-Health Systems—A Comprehensive Review.* Symmetry; Multidisciplinary Digital Publishing Institute.

7. Mehta, D.D., Zanartu, M., Feng, S.W., Cheyne, H.A., Hillman, R.E. (2012). Mobile voice health monitoring using a wearable accelerometer sensor and a smartphone platform. *IEEE Transactions on Biomedical Engineering*, 59(11), 3090–3096.

8. Gao, H., Duan, X., Guo, X., Huang, A., Jiao, B. (2013). Design and tests of a smartphone-based multi-lead ECG monitoring system. *2013 35th Annual International Conference of the IEEE Engineering in Medicine and Biology Society (EMBC)*, 2267–2270. http://dx.doi.org/10.1109/EMBC.2013.6609989

9. Chen, W., Chen, Z., & Cui, F. (2019, May 20). Collaborative and secure transmission of medical data applied to mobile healthcare. *Biomedical Engineering Online*, BioMed Central. https://doi.org/10.1186/s12 938-019-0674-x

10. Miller, T. (2019). Explanation in artificial intelligence: Insights from the social sciences. *Artificial Intelligence*, 267, 1–38. https://doi.org/10.1016/ j.artint.2018.07.007

Chapter 11

Design challenges in cyber-physical systems

An overview of privacy threats in ubiquitous computing

*Manivel Kandasamy, Raju Shanmugam, Vishva Patel,
Umika Patel, Ananya Bulchandani, and Aditya Bhardwaj*
Unitedworld Institute of Technology, Karnavati University, Gandhinagar,
Gujarat, India

11.1 INTRODUCTION

11.1.1 What are cyber-physical systems?

The cyber-physical system (CPS) is the type of bodily gadget that combines the physical and the computational components aiming for the brand-new magnificence of shrewd systems for attaining high overall performance. The cyber bodily systems are designed to integrate the communique technologies with the physical structures that could engage with people without difficulty via new strategies [1].

The important features of CPS are its potential to feel, actuate, and compute. Sensing refers back to the potential of the device to accumulate information from its environment and the usage of a diffusion of sensors, inclusive of cameras, microphones, and temperature sensors.

Actuation refers to a machine's ability to use actuators, including vehicles, valves, and switches, and computing refers to the device's potential to the method and examine facts, the usage of algorithms, and choice-making processes [2].

Examples of CPS consist of medical devices and structures [3], transportation systems and intelligent highways, industrial automation, and clever homes. For instance, in transportation systems, the cyber bodily machine is used to combine the sensors, communicate structures and manage the algorithms to manipulate the visitors to go with the flow [4].

11.1.2 Challenges and opportunities in cyber-physical systems

Research in the field of CPS can be accelerated by identifying needs, challenges, and opportunities in several industrial sectors and by encouraging multidisciplinary collaborative research between academia and industry [5]. The objective is to expand new systems technology and engineering methods for building high-self-assurance structures wherein cyber and physical designs are compatible, synergistic, and integrated with any respect scales. The core technological know-how and generation required to help the CPS vision for destiny economy.

CPS are vulnerable to cyber-attack, and the consequences could be severe. Security must be considered at every stage of CPS design and implementation, from hardware to software communication protocols and network infrastructure. As improper security leads to unsafe systems, CPS is expected to operate in real-time and interact with the physical world, which will then create a safe version out of it. As unsafe CPS could lead to an unsafe system, privacy is another challenge in CPS. CPS can collect and transmit sensitive data, which raises privacy concerns [6].

CPS is inherently complex, which creates additional challenges. Designing, implementing, and maintaining CPS requires a multidisciplinary approach. Standard and open interfaces can help address these challenges as well as model-driven techniques and middleware solutions. Fault tolerance and redundancy can help ensure that CPS continues to operate even in the face of failures [7].

In conclusion, addressing the challenges in CPS requires a multidisciplinary approach, involving experts from multiple fields. It also requires the development of new tools, techniques, and standards to support the design, implementation, and operation of CPS. By addressing these challenges, we can ensure that CPS is secure, safe, interoperable, reliable, and respectful of users' privacy.

11.1.3 Industrial automation and cyber-physical systems

Industrial automation and cyber-physical systems are interrelated as they both involve the use of advanced technologies. Nowadays, many industrial sectors as well are using the CPS for the betterment and more efficiency to improve their performance. Industrial automation and CPS both contain widespread cybersecurity dangers, as they are noticeably linked and rely on networked technology. As such, it is important to make certain that these systems are designed with security and thoughts, and they are often up to date and maintained to defend the cybersecurity threats [5, 8].

The integration of industrial automation with cyber-physical systems has led to the emergence of a new generation of intelligent manufacturing systems, often referred to as industry 4.0 [9]. This integration allows for the creation of intelligent, autonomous manufacturing systems that can sense, analyse, and adapt to the changes in the production environment. The data can be analysed using artificial intelligence using algorithms to identify potential issues and improve product efficiency, which conducts significant cost savings, increased productivity, and improved product quality.

11.2 ARCHITECTURE OF CYBER-PHYSICAL SYSTEMS

11.2.1 Overview of cyber-physical systems architecture

Cyber-physical system contains two main layers. They are the physical layer, which includes sensors and actuators, and cyber layer, which includes computing and communication systems. CPS system includes variables that represent data received by sensors and control variables that represent control signals. Controllers calculate the distance between the values of the process variables and the corresponding control actuator.

CPS may consist of multiple sensors and actuators and include various combinations of key functions. The characteristics to be used in a CPS depends on the field of application. A CPS architecture can be divided into various levels [10].

11.2.1.1 Sensing module

Sensing module collects data through sensors and sends them to the data management module. It supports multiple networks. For example, in HCPS, using BAN, sensors attached with patients are equipped with sensing module nodes to enable real-time control.

11.2.1.2 Data management module

Data management module consists of the computational devices and storage media. This provides the heterogeneous data processing such as normalisation, noise reduction, data storage, and other similar functions. DMM collects the sensed data and forwards it to service aware modules of next generation Internet.

11.2.1.3 Next generation Internet

Unlike the current Internet architecture where routing protocols find a single (the best) path between a source and destination, future Internet routing protocols will need to present applications with a choice of path. Next generation Internet is the ability for applications to select the path, or paths that their packets take between the source and destination.

11.2.1.4 Service aware module

Service aware module provides functions of the whole system. After it receives signals, it recognises and sends data to services available.

11.2.1.5 Application module

A number of services are deployed and interact with NGI in application module. Simultaneously, information is getting saved on secured database. Database is maintained at local storage and on cloud platforms at the same time in order to keep data safe. This saved data over the cloud system can be accessed from anywhere followed by authenticated access.

11.2.1.6 Sensors and actuators

Sensor and actuator modules are electronic devices that interact with physical environment. An actuator is a physical device that receives commands from application module and executes.

11.2.2 Physical architecture

The physical architecture network of a CPS as shown in Figure 11.1 refers to the way in which these physical components are interconnected and communicate with each other. The physical architecture network of a CPS typically includes sensors that gather data about the physical world, such as temperature, pressure, and position. These sensors may be connected to microcontrollers or other computing devices that process the data and make decisions about how to actuate the system in response. Actuators, such as motors, valves, or pumps, then use this information to control the physical components of the system. In addition to these physical components, the physical architecture network of a CPS may also include communication networks that enable the various components of the system to exchange data and commands. This may include wired or wireless communication networks, such as Ethernet, Wi-Fi, or Bluetooth, as well as protocols for managing the flow of data between the different components. The physical architecture network of a CPS must be carefully designed to ensure that the system operates reliably and efficiently. This may involve optimising the placement of sensors and actuators, selecting appropriate communication protocols, and implementing redundancy and failover mechanisms to ensure that the system can continue to operate even in the event of component failures or other disruptions.

11.2.3 Network architecture

The network architecture of a CPS is a critical element in its design, as it determines how the different components of the system will communicate and interact with each other. The key elements of the network architecture of CPS are as follows:

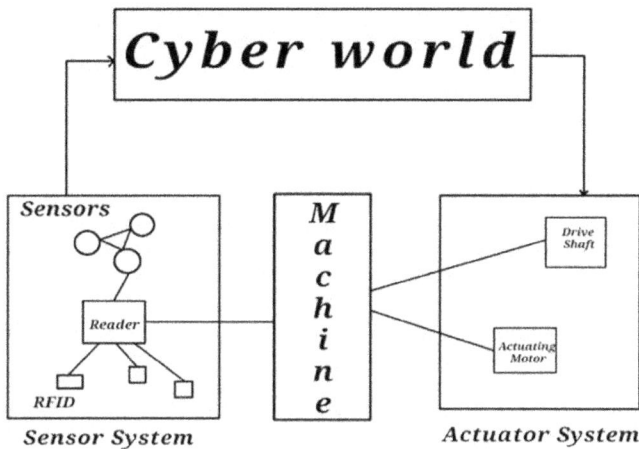

Figure 11.1   Physical architecture network.

1. Network topology

    The network topology refers to the physical or logical arrangement
    of the devices or components in a CPS. Depending on the application
    requirements, different topologies can be used, such as star, mesh, bus,
    or ring topologies.

2. Communication protocols

    Communication protocols define the rules and format for data exchange
    between the components of a CPS. Some common communication
    protocols used in CPS include Modbus, CAN, ZigBee, and OPC-UA.

3. Sensors and actuators

    Sensors are devices that collect data about the physical environment,
    while actuators are devices that allow the CPS to control physical
    systems. These devices can be connected to the network using various
    protocols, such as Ethernet, Wi-Fi, or Bluetooth.

4. Middleware

    Middleware provides a layer of abstraction between the hardware
    and software components of a CPS, allowing them to communicate
    with each other without requiring detailed knowledge of each other's

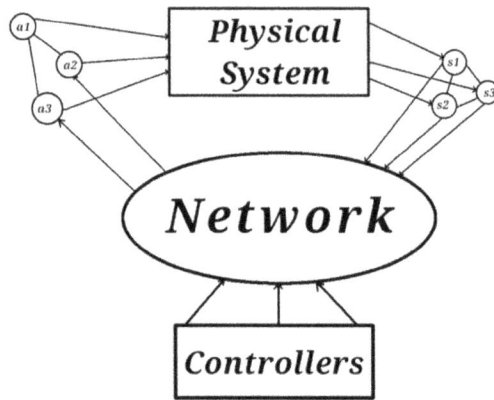

Figure 11.2   Network architecture of CPS.

implementation. Examples of middleware used in CPS include message queuing telemetry transport (MQTT) and data distribution service (DDS).

5. Cloud integration

   In many CPS applications, cloud computing can be used to store, process, and analyse large amounts of data generated by the system. Cloud integration involves connecting the CPS to cloud services, which can be done using various communication protocols and middleware.

6. Security

   Due to the sensitive nature of many CPS applications, security is a crucial aspect of their network architecture. This includes measures such as encryption, access control, and intrusion detection to protect against unauthorised access and malicious attacks.

Overall, the network architecture of a CPS, as shown in Figure 11.2, must be carefully designed to ensure that it meets the application requirements while providing robust, reliable, and secure communication between its various components.

11.2.4 Data management and security

Data management involves organising, storing, and maintaining data in a structured and efficient manner to enable effective use and retrieval of information. Data security involves protecting data from unauthorised access, use, or disclosure [11].

11.2.4.1 Data management

Data management involves several processes, including data capture, storage, integration, processing, analysis, and sharing. Data capture involves collecting data from various sources, while storage involves preserving the data for future use. Integration involves combining different data sets to create a unified view, and processing involves transforming data into a usable form. Data analysis involves using various tools and techniques to gain insights and knowledge from data, while data sharing involves providing access to data to users who need it. Effective data management requires a robust infrastructure, including hardware, software, and network components. It is essential for businesses, organisations, and individuals to ensure the privacy, confidentiality, and integrity of sensitive and valuable data.

Key concepts of data management, as shown in Figure 11.3, are discussed below:

1. Data governance: It refers to the process of defining and implementing policies, procedures, and controls for managing and protecting data assets.

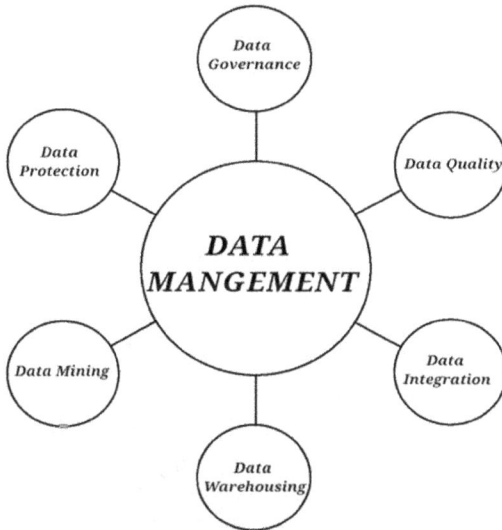

Figure 11.3   Network data management.

2. Data quality: It is the degree to which data is accurate, consistent, complete, and relevant for its intended purpose.
3. Data integration: It involves combining data from multiple sources to create a unified view of the data.
4. Data warehousing: It refers to the process of collecting, storing, and managing large amounts of data in a central repository.
5. Data mining: It involves analysing data to discover patterns, relationships, and insights that can be used for decision-making.
6. Data protection: It includes the measures taken to safeguard data from unauthorised access, use, disclosure, alteration, destruction, or theft.

11.2.4.2 Data security

Data security is the practice of protecting digital information from unauthorised access, corruption, or theft throughout its entire lifecycle. It involves the implementation of various measures to ensure that data is kept safe and secure, both while it is stored and when it is being transmitted. Data security is important because it helps to prevent sensitive information from falling into the wrong hands, which can have serious consequences for both individuals and organisations. For example, if personal information such as social security numbers or credit card details are stolen, they can be used for identity theft or fraud [12].

There are various methods that can be used to improve data security, including encryption, firewalls, access controls, and monitoring systems as shown in Figure 11.4. These measures can help to protect against cyberattacks and other forms of unauthorised access to data. There are four types of data security:

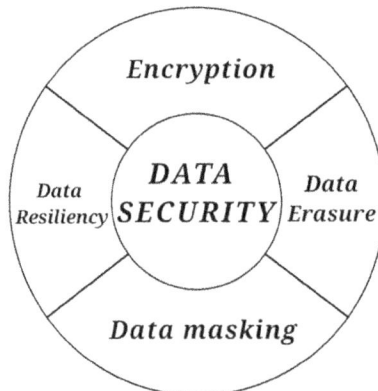

Figure 11.4   Data security.

1. Encryption

    A computer algorithm transforms text characters into an unreadable format via encryption keys. Only authorised users with the proper corresponding keys can unlock and access the information.

2. Data erasure

    Data erasure uses software to completely overwrite data on any storage device. It verifies that the data is unrecoverable. There will be occasions in which organisations no longer require data and need it permanently removed from their systems. Data erasure is an effective data security management technique that removes liability and the chance of a data breach occurring.

3. Data masking

    Data masking enables an organisation to hide data by obscuring and replacing specific letters or numbers. This process is a form of encryption that renders the data useless should a hacker intercept it. The original message can only be uncovered by someone who has the code to decrypt or replace the masked characters. By masking data, organisations can allow teams to develop applications or train people using real data. It masks personally identifiable information (PII) so that development can occur in environments that are compliant.

4. Data resiliency

    Resiliency is determined by how well an organisation endures or recovers from any type of failure – from hardware problems to power shortages and other events that affect data availability. Organisations can mitigate the risk of accidental destruction or loss of data by creating backups or copies of their data.

11.3 DESIGN CHALLENGES FOR CYBER-PHYSICAL SYSTEMS

11.3.1 Real-time performance

The ability of a real-time operating system (or any programme, for that matter) to respond quickly to events is known as real-time performance. Real-time indicates that a software response to one event must occur before a second occurrence that occurs independently of the first.

The Oxford Dictionary of Computing gives the following definition of a real-time system: Any system in which the time at which output is produced is significant. This is usually because the input corresponds to some movement in the physical world, and the output has to relate to that same movement. The lag from input time to output time must be sufficiently small for acceptable timelines.

The events to which real-time systems must respond are typically made known to them as interrupts, and they are typically designed to be reactive in nature. When the processor detects an interrupt, it takes specific actions and runs instructions that were created to respond to the situation.

Most of the time, the CPU is already carrying out some instructions when it first notices the interrupt. This processing needs to be "interrupted," and after the crucial real-time interruption response is finished, it can be resumed.

Real-time operating systems must perform additional tasks in addition to interrupt handling. They must also plan and control how application software threads are executed. Requests from threads for scheduling, message passing, resource allocation, and many other services are handled by the real-time operating systems. Most of the time, services must be completed fast to allow the thread to finish its assigned task before the next interrupt occurs.

Practitioners in the field of real-time computer systems design often differentiate between hard and soft real-time systems [13]:

- Hard real-time systems (HRTSs): systems in which replies must be delivered within a specific time frame.
- Soft real-time systems (SRTS): systems in which response times are critical yet the system will continue to function properly if deadlines are periodically missed. Soft real-time systems can be separated from interactive systems that lack explicit deadlines.

Disadvantages of real-time operating system

a.   Multitasking

When it comes to multitasking, a real-time operating system can focus on the target application but not on other tasks. They are just intended to do a few functions. As a result, it is not recommended for multitasking systems.

b.   Driver request

Signal disruptions are unavoidable in real-time operating systems. As a result, in order to attain a given speed, the relevant drivers must be installed on your computer. When an interrupt occurs, the Real-time operating system can reply fast by using drivers.

c.   **Programme crashes**

By employing a real-time operating system, frequent programme crashes are possible. Unlike traditional operating systems, real-time operating system cannot effectively divide storage domains. As a result, processes have difficulty dealing with them.

d.   **Task focus**

Real-time operating system concentrates on a single application at a time. This is done largely to ensure accuracy and reduce errors. All other low-priority applications must be postponed. There is no time limit for waiting. Underneath the real-time operating system interface lie complex, difficult algorithms. Ordinary folks find it challenging to construct these algorithms. They can only be written and understood by experienced developers.

11.3.2 Hardware and software integration

The act of designing software that integrates with one or more pieces of real hardware to create a single usable system is known as hardware integration (also known as hardware–software integration or system integration) time to market. Instead of commencing firmware and software development and testing after the hardware design is available, Windows forces hardware and software engineers to work concurrently. Complex hardware and software methodologies necessitate not just verification of each individual subsystem's correctness, but also co-verification of their right interactions after integration.

Because the industry has to further integrate hardware and software implementation and debug, design teams frequently use an equal or greater number of software engineers in proportion to hardware engineers.

11.3.3 Fault tolerance and reliability

Fault tolerance is the ability to meet requirements notwithstanding the occurrence of system flaws. A more complicated control system necessitates effective fault tolerance. For cyber-physical systems, fault tolerance has long been a serious concern. It frequently takes the form of huge redundancy in cyber-physical systems, with voting used to disguise component failures.

For instance, the system is in a state which allows for a lowered level of fault tolerance. Such a lowered level can be useful in several ways:

• A lower computational load enables the adoption of power management strategies to reduce the amount of energy consumed by the computer system. Given the energy constraints of many cyber-physical applications, this is frequently quite useful.

- Lowering the quantity of energy consumed reduces the amount of heat generated; processors can run cooler under lighter loads. Running cooler leads to increased reliability because temperature is strongly connected with the rate at which CPUs age.
- Lowering the computational load associated with critical operations allows the system to allocate more resources to non-critical tasks, improving control quality.

11.3.4 Scalability and interoperability

Interoperability refers to the ability of systems to work together, or inter-operate. Initially, the phrase referred to services in information and communication technology and systems engineering that enabled smooth information interchange. A broader definition includes social, political, and organisational aspects that influence system performance. Interoperability is the task of creating coherent services for systems when individual system components are technically diverse and governed by various software systems.

The ability of a system to vary its performance and cost in response to changes in application and system processing demands is referred to as scalability. Examples include how effectively a hardware system handles expanding user counts, how well a database handles increasing query volumes, and how well an operating system works on various types of hardware. When it comes to technology and software, rapidly growing businesses should pay special attention to scalability.

11.3.5 Security and privacy

The very structure of CPSs allows for both cyber and physical assault paths, significantly expanding the adversary's options. Different sets of vulnerabilities on the cyber and physical sides do not simply add up; they increase. Physical access to a cyber system allows for attacks that would not be possible otherwise. The addition of a networked cyber component to a physical system enhances the system's complexity, the scope of what can be attacked, and the distance from where the attack can be carried out [14].

Separate attack pathways may be fully secured in only one domain, but only areas of the system where both domains are protected at the same time are truly protected. Simultaneously, defence in either the cyber or physical component can be employed to protect the other component in more ways than a pure cyber or physical system could. Attack points for security and privacy in CPSs can be found at device interfaces, on the devices themselves, in the infrastructure that supports them, from the Internet, and even from malicious individuals.

Attackers may use the ambiguities of vulnerable communication protocols to launch a cross-interface attack. They may compromise a component by exploiting security holes in poorly implemented application programming interfaces. Conversely, they may exploit trust connections between peer devices or between devices and the infrastructures, clients, and users with whom they communicate. Each of these vulnerability sites must be protected by security measures and regarded as potentially compromised system components by other components as shown in Figure 11.5.

Inadequate deterrent, detection, delay, and authentication make the system vulnerable to a cyber-attack, which could result in disastrous physical impacts. We can address the lack of physical protections by implementing measures such as cameras and stronger physical barriers. Adding alarms would also improve detection and response time. To improve authentication, the system should require users to have unique identifiers and passwords, so that even if someone connected a laptop directly into the PLC, he or she would be unable to access the system without first logging in. Barriers and

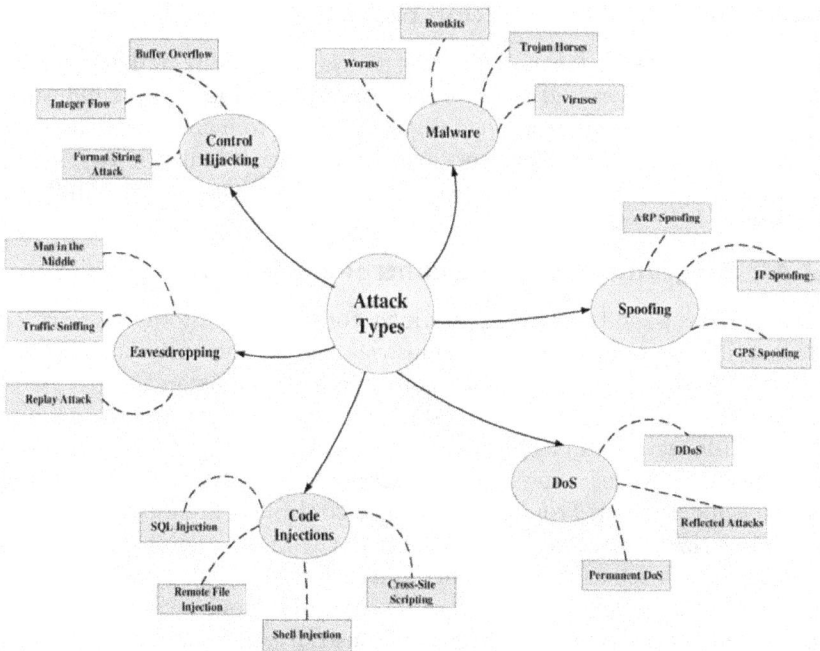

Figure 11.5 Cyber-physical system attacks.

identities would also lengthen the time it takes to utilise the system, allowing authorities more time to react.

Privacy issues cannot be fully comprehended without a detailed understanding of the entire system and its interconnections. One issue with CPSs is that links to bigger networks like the Internet are not always visible.

According to Groopman and Etlinger [9], customers are increasingly concerned about the data collected about them and how it will be utilised. Data acquired is potentially shared silently, especially in the age of IoT [15]. Data had to be manually entered into a computer previously. People who understand that their devices are connected to the Internet often do not understand the privacy implications.

Wearable gadgets may interact with collection locations in stores, restaurants, along roads, or everywhere we travel, and these collection stations may be undetectable. Collecting points may compel nearby devices to expose their identities and connect to the Internet via the collection point as a go-between. The infamous Fitbit sexual activity data-sharing controversy likewise had unclear limits and unforeseen consequences of sharing.

11.4 SUPPORTING TECHNOLOGIES FOR INDUSTRIAL AUTOMATION

11.4.1 Industrial automation technologies

Industrial automation technology refers to a variety of systems that comprise software and hardware that are used to automate and control industrial processes. Such technologies are designed in order to increase productivity, reduce costs and improve efficiency in a wide range of industries.

Some significant technologies used in industrial automation are as follows.

11.4.2 Programmable logic controllers (PLCs)

PLCs are the digital technologies which control and automate industrial processes. It describes a field of industrial automation solutions that can perform multiple functions and are controllable via commands delivered by the means of entering computer code in the systems. They are designed to be more adjustable than fixed tools. It allows customisation and adjustment of the manufacturing equipment in accordance with the requirements of each specific product. Programmable logic controllers are modular devices of various size that include a microprocessor with the appropriate number ranging from dozen to thousands of input and output [16]. They are used to interconnect different kinds of industrial solutions in one network, enabling automated control and monitoring of industrial machinery and processes. The key role of PLCs in industrial automation can be broken down into the following areas:

a. **Control:** PLCs are used to control a wide range of industrial processes, including machine and system control, process control and motion control. PLCs can be programmed to monitor inputs from sensors, and other devices and output control signals to control actuators, motors, and other industrial equipment.

b. **Communication:** PLCs are designed to communicate with other industrial equipment and systems including sensors, HMIs, SCADA systems and other PLCs. This allows for coordinated control of complex systems and improved data collection and analysis.

c. **Data acquisition:** PLCs can be programmed to collect data from sensors and other devices and store it in memory or transmit it to other systems for analysis. This data can be used to monitor the performance of industrial equipment and systems, identify potential issues, and improve the efficiency of manufacturing and production processes.

d. **Safety:** PLCs are used to monitor and control safety systems in industrial automation technologies. PLCs can be programmed to monitor emergency stop buttons, safety interlocks, and other safety devices to ensure safe operation of industrial equipment and systems.

11.4.3 Human machine interfaces (HMIs)

HMIs are graphical user interfaces which allow humans to interact and control automated systems. They are a critical component of industrial automation systems, providing operators with real-time information about the status and performance of machines and processes [17]. The role of HMI in industrial automation technologies can be broken down into the following areas:

a. **Monitoring:** HMIs provide operators with real-time information about the status and performance of machines and processes. This information includes data on machine status, stating that whether the machine is running or stopped, as well as process variables such as temperature, pressure, and flow rate. Operators can use this information to monitor the performance of machines and processes and identify potential issues before they become significant problems.

b. **Control:** HMIs provide operators with the ability to control the operation of machines and processes. This includes the ability to start and stop machines, adjust process variables and troubleshoot issues. HMIs also provide operators with the access to machine and process settings, such as operational modes and setpoints, enabling them to make adjustments to optimise performance.

c. **Alarms and notification:** HMIs can be programmed to generate alarms and notification when certain events occur such as, when a machine is out of tolerance or when a process variable exceeds a set limit. Alarms and notifications can be displayed on the HMI screen, sent via email or text message or broadcast to other systems.

d. **Customisation**: HMIs can be customised to meet specific industrial automation needs. This includes the ability to create custom displays, alerts, and notifications, as well as the ability to integrate with the other industrial automation systems.

11.4.4 Supervisory control and data acquisition (SCADA) systems

These systems are used to monitor and control industrial processes from a central location [18]. SCADA is a term for all kinds of complex industrial control systems that use a combination of components such as computers, graphical user interfaces, and network data communication to provide a high level of automated controls and monitoring of processes. SCADA systems play a critical role in industrial automation. They provide operators to control and monitor industrial processes, access real-time information, and generate alerts and notification when needed. SCADA systems also enable data acquisition and historical data collection and analysis, allowing operators to optimise processes and improve overall efficiency. Effective SCADA design is critical to ensure the efficient and safe operation of industrial automation systems.

11.4.5 DCS (distributed control system)

It is a digital automated industrial control system that uses geographically distributed control loops throughout a machine or control area. The agenda of DCS is to control industrial processes to increase their safety, cost effectiveness, and reliability.

DCS systems play a critical role in ensuring the safe operation of industrial processes. They are designed to detect and prevent hazardous conditions, such as overpressure, overtemperature, or equipment failure. DCS systems can also initiate emergency shutdowns and other safety measures to protect the equipment.

They also ensure the integration of hardware and software with redundancy providing a high degree of system availability and reliability. This redundancy ensures that critical processes continue to operate even if one or more components of the system fails.

a. **Machine vision**: Machine vision systems use cameras and image processing software to perform tasks such as quality control, defect detection, part identification, inspection, and measurement. It is commonly used in industries, such as manufacturing, automotive, and food processing, among others. The role of machine vision in industrial automation can be broken down by the following ways:

b. **Inspection:** Machine vision is used to inspect products and components for defects and anomalies. This can include surface defects, missing or damaged components, or incorrect assembly. Machine vision use high-resolution cameras and specialised software to analyse images of the products and compare them to predetermined specifications. If a defect is detected, the machine vision can generate an alert or initiate a corrective action.

c. **Guidance:** Machine vision is used to guide robotic systems or other automation equipment. This can include identifying the location of parts or components, verifying correct placement, and ensuring accurate alignment. Machine vision systems can also provide real-time feedback to robotic systems, allowing them to adjust their position or orientation to improve accuracy.

d. **Measurement:** Machine vision is used to measure dimensions or features of products and components. This can include length, width, height, or other features such as hole sizes or distances between objects. Machine vision use specialised software to calculate these measurements and compare them to predetermined specifications.

e. **Sorting:** Machine vision is used to sort products or components based on specific criteria. This can include size, shape, colour, or other features. Machine vision can quickly and accurately identify these criteria and sort the products into different categories.

f. **Traceability:** Machine vision is used to track products or components through the manufacturing process. This can include identifying individual components and tracking them through the assembly process or tracking finished products through the supply chain. Machine vision can provide real-time feedback on the location and status of products enabling manufacturers to optimise their processes and improve efficiency.

11.4.6 Industrial Internet of Things (IIOT)

IIOT refers to the use of sensors and other devices to collect data and transmit it over a network, allowing for real-time monitoring and control of industrial processes. In other words, IIOT refers to the extended use of IoT in industrial automation solutions, describing the ecosystem of sensors, machines, robotic devices, and other instruments connected together, communicating and exchanging data with both internal and external software automation systems. IIOT is widely used across the industries, enabling improved productivity, efficiency, and analytics with a combination of innovative technologies that are fuelling the new generation of industrial automation [19, 20].

Overall industrial automation technology has transformed the manufacturing industry by making it faster, more efficient, and safer. It has reduced the possibilities of human error and increased productivity while also enabling manufacturers to produce high-quality products.

11.4.7 Industrial networking technologies

Industrial networking technologies refer to the communication technologies used in industrial automation systems [21]. These technologies provide a means of data transmission between different components of an automation system: sensors, actuators, controllers, and HMIs. Some industrial networking technologies include the following:

Ethernet: Ethernet is widely used networking technology in industrial automation systems. It provides high-speed data transmission and is used for both local and wide area networks. Ethernet is used to provide a high-speed, reliable communication network for connecting devices, sensors, and control systems. It is used for a variety of applications including process control, motion control, data acquisition, and supervisory control and data acquisition (SCADA).

Profibus: Profibus is an industrial networking technology that is used to connect field devices to automation systems. It provides a fast and reliable means of communication between devices and supports a wide range of applications. It is one of the well-known and widely implemented open field networks. These networks are mainly used in process automation and factory automation fields. It is most suitable for complex communication tasks and time critical applications. There are three versions of Profibus namely, Profibus-DP (Decentralised Periphery), Profibus-PA (Process Automation) and Profibus-FMS (Fieldbus Message Specifications).

Profibus-DP is an open fieldbus communication standard that utilises master/slave communication between network devices. It uses RS485 or fibre optic transmission technologies as physical layer media. It is mainly used to provide communication between controllers and distributed input and output at device level.

Profibus-PA is specially designed for process automation. Profibus-PA networks are recommended for use in intrinsically safe areas. These network permit sensors, actuators, and controllers to connect to a single common bus, which provide data communication and power over bus.

Profibus-FMS is a multi-master or peer-to-peer messaging format, which allows master units to communicate with one another. It is a general-purpose solution that performs communication tasks in control level, especially in cell sublevel to facilitate communication between master PCs.

Most commonly FMS and DP are used simultaneously in COMBI mode in situations where PLC is being used in conjunction with a PC. In such a case, the primary master communicates with the secondary master via FMS while DP transfers control data on same network to input and output devices.

Modbus: Modbus is a serial communication protocol that is commonly used in industrial automation systems. It is often used to connect devices such as PLCs and HMIs to sensors and actuators. It is an open system protocol that can run on a variety of physical layers. It is the most widely used protocol in industrial control applications. It is a serial communication technique which provides a master/slave relationship to communicate between devices connected on network. It can be implemented on any transmission medium, but most commonly used with RS232 and RS485. Serial Modbus with RS232 and RS485 facilitates the connection of Modbus devices to the controller such as PLC in a bus structure. It can communicate between one master and a number of slaves up to 247 with a data transmission rate of 19.2k bits/sec. The newer version of Modbus TCP/IP uses Ethernet as a physical layer that facilitates the data exchange between PLC in different networks. Irrespective of the type of physical network, it facilitates the method of access and control of one device by another.

ControlNet: ControlNet is a proprietary networking technology developed by Rockwell automation. It provides a high-speed, deterministic network for industrial automation applications. It is an open control network, which uses Common Industrial Protocol (CIP) in order to combine the functionality of peer-to-peer network and an input and output network by providing high-speed performance. This network is the combination of Data Highway Plus (DH+) and remote input and output. It is used for real-time data transfer of time critical as well as non-time critical data between input and output or processors on the same network. It can communicate up to a maximum of 99 nodes with a data transfer of five million bits per second. It was designed to be used on both device and field level of industrial automation systems. It provides media and communication redundancy to all the nodes of the network.

Device Net: Device Net is another proprietary networking technology developed by Rockwell automation. It is used for connecting sensors and actuators to automation systems and provides low-cost solution for industrial networking. It is an open level device network based on CAN technology. It is designed to interface field-level devices (such as sensors, switches, barcode readers, panel displays, etc.) with higher level controller such as PLC with a unique adoption of basic CAN protocol. It can support up to 64 nodes and supporting up to 2048 total devices. It can reduce the network cost by integrating all devices on a four-wire cable that carries both data and power conductors. The power on the network allows the devices to be powered up directly from the network and hence it reduces the physical connection points. This network is popularly used in automotive and semiconductor industries.

Overall industrial networking technologies play a critical role in modern industrial automation systems. They enable devices to communicate and work together, providing efficient and reliable control of industrial processes.

11.4.8 Industrial data management technologies

Industrial data management technologies refer to the tools, systems, and processes used by companies to collect, store, analyse, and use data generated from their industrial processes. These technologies are crucial for managing and optimising industrial operations, reducing costs, improving efficiency, and increasing productivity [22].

The following are some key industrial data management technologies:

1. **Data acquisition systems (DAS):** DAS are used to collect data from sensors and equipment in industrial processes. They can collect data on parameters such as temperature, pressure, flow rate, and energy usage. DAS can be software- or hardware-based and may use various protocols to communicate with sensors and other equipment.
2. **Supervisory control and data acquisition (SCADA) systems:** SCADA systems are used to control and monitor industrial processes. They collect data from sensors and provide real-time feedback to operators. SCADA systems are commonly used in the energy, water and transportation industries.
3. **Manufacturing execution systems (MES):** MES are used to manage and control manufacturing processes. They collect data from various sources such as DAS and SCADA systems, and provide real time information to help managers to make decisions. MES can also integrate with other systems such as enterprise resource planning (ERP) systems, to provide a comprehensive view of operations.
4. **Enterprise resource planning (ERP) systems:** ERP systems are used to manage business processes including inventory, purchasing, and accounting. They collect data from various sources such as MES and SCADA systems, and provide a unified view of the business operations. ERP systems can help managers make informed decisions by providing real-time information on inventory levels, production schedules, and other key performance indicators.
5. **Data historians:** Data historians are used to store and retrieve large amount of historical data generated by industrial processes. They are commonly used in energy, oil, gas, and chemical industries. Data

historians can provide insight into past performance, identify trends, and support predictive analytics.

6. **Big data analytics**: Big data analytics is the process of analysing large volumes of data to identify patterns and insights. It is used in manufacturing, energy, and transportation to optimise operations and improve efficiency. Big data analytics can provide insight into factors affecting production and can help predict maintenance needs.

7. **Cloud computing**: Cloud computing is used to store and process data on industrial data on remote servers. It provides flexibility and scalability, making it easier for the companies to manage large volumes of data. Cloud-based systems can also provide real-time access to data from any location, enabling remote monitoring, and collaboration.

8. **Internet of Things (IoT)**: IoT refers to the network of physical devices, vehicles, and other objects that are embedded with sensors, software, and network connectivity. IoT is used to collect data from industrial equipment and provide real-time feedback to operators. IoT can also enable predictive maintenance by identifying equipment issues before they cause downtime.

These technologies are essential for managing and optimising industrial operations. By collecting and analysing data, companies can find areas for improvement, increase efficiency, and reduce cost. Effective data management requires robust systems, reliable sensors, and skilled professionals to analyse and interpret data. As industrial processes become increasingly complex and data-intensive, the importance of effective data management will continue to grow.

11.4.9 Industrial robotics technologies

Industrial robotic technologies refer to the use of robotic systems in industrial applications to automate and optimise manufacturing and production processes. These technologies are designed to perform repetitive and dangerous tasks with greater speed, accuracy, and efficiency than human workers [23].

There are several robotic technologies in industrial sector including the following:

Cartesian robots: Cartesian robots are designed with three linear axes that move in a straight line. They are commonly used in pick and place applications and material handling. The key role of Cartesian robots in industrial robotic technology can be broken down into the following ways:

High precision: Cartesian robots are designed with high precision and accuracy, making them ideal for applications that require exact positioning of objects or tools. This precision is achieved through the use of linear actuators and linear motion guides that ensure smooth and accurate movement in all three dimensions.

Fixed workspace: Cartesian robots are typically used in applications where workspace is fixed, such as assembly lines or production cells. They are designed to move within a defined area, which makes them ideal for applications that require repeated movements within a fixed space

High payload capacity: Cartesian robots are capable of carrying heavy payloads, which makes them ideal for application that require the manipulation of large or heavy objects. This payload capacity is achieved through the use of strong and durable materials, such as steel or aluminium.

Versatility: Cartesian robots are highly versatile and can be used in a wide range of industrial applications. They can be equipped with different end-effectors, such as grippers, suction cups, or cutting tools, which allows them to perform a variety of tasks.

Easy programming: Cartesian robots are easy to program, thanks to their simple Cartesian coordinate system. Programming can be done through graphical interfaces or software, which simplifies the programming process and reduces the need for specialised programming skills.

Safety: Cartesian robots are equipped with safety features, such as limit switches, emergency stop buttons, and safety sensors, which ensure safe operation in industrial environments. These safety features help to prevent accidents and reduce the risk of injury to operators.

Articulated robots: Articulated robots are most commonly used industrial robots that are designed with multiple joints that allow for a wide range of motion and flexibility. They are commonly used in welding, material handling, and assembly applications. The role of articulated robots in industrial robotic technology is to provide high flexibility and dexterity in a large workspace. Their versatility, high payload capacity, and easy programming make them ideal for a wide range of industrial applications, including welding, material handling, and assembly applications. Effective implementation and management of articulated robots are essential for ensuring the efficient and reliable operation of industrial automation systems.

SCARA robots: SCARA (selective compliance assembly robot arm) robots are designed with two parallel arms that move in the X and Y axis. They are commonly used in packaging, assembly, and material handling applications. The key role of SCARA robots in industrial robotic technology can be broken down in the following ways:

High speed and precision: SCARA robots are designed to move quickly and accurately, which makes them ideal for applications that require high speed and precision. Their vertical axis provides stability, while their flexible horizontal arm allows them to move in a circular or linear path.

Pick and place applications: SCARA robots are often used for pick and place applications, where they can quickly and accurately pick up parts or products and place them in a specific location. This is useful in industries such as electronics manufacturing and packaging.

Assembly applications: SCARA robots are also used in assembly applications, where they can be programmed to fit, join, or assemble parts together. This is useful in industries such as automotive and aerospace.

Compact design: SCARA robots have a compact design, which makes them ideal for use in environments where space is limited. Their small footprint means that they can be used in tight spaces, such as on assembly lines or in clean rooms.

Easy programming: SCARA robots are relatively easy to program, thanks to their simple design and the availability for user friendly software. This simplifies the programming process and reduces the need of specialised programming skills.

Cost-effective: SCARA robots are typically more affordable than other types of industrial robots, which makes them a popular choice for small- and medium-sized businesses. Their cost-effectiveness, combined with their versatility and ease of use, make them an attractive option for businesses looking to automate their processes.

DELTA robots: Delta robots are designed with three arms that move in triangular motion. They are commonly used in food and beverages packaging applications. The key role of delta robots in industrial robotic technologies are broken down in the following ways:

1. **High speed and precision:** Delta robots are designed to move very quickly and precisely making them ideal for applications that require fast, accurate movements. Their triangular configuration allows them to achieve a high level of stiffness and accuracy, which makes them well suited for tasks, such as pick and place, packaging, and assembly.
2. **Light payloads:** Delta robots are typically used for applications that involve light payloads, such as small parts or food products. Their speed and precision make them well suited for these types of applications, which require fast and accurate movements.
3. **Small footprint:** Delta robots have a small footprint and take up a very little floor space. This makes them ideal for use in applications where space is limited, such as in the food and beverage industry or in clean room environments.
4. **Flexible automation:** Delta robots can be easily programmed to perform a variety of tasks and can be reprogrammed quickly to

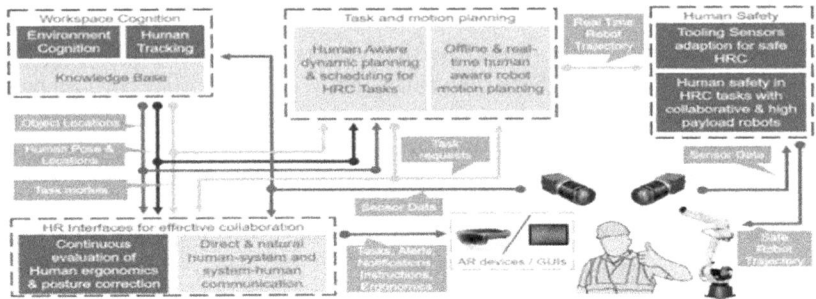

Figure 11.6   Functional block diagram of industrial robotic technologies.

adapt to changing production needs. This flexibility makes them ideal for use in industries that require a high degree of customisation, such as the automotive or electronics industries.

5. **Easy integration:** Delta robots can be easily integrated into existing production lines, which reduces downtime and increases productivity. They can also be combined with other automation technologies, such as vision systems or conveyor belts, to create a fully automated production line.

Collaborative robots: Collaborative robots are designed to work alongside human workers. They are equipped with safety measures such as sensors and cameras that allow them to work safely with human workers. They are commonly used in assembly and material handling applications. The role of collaborative robots in industrial robotic technology is to assist humans in programming tasks that are too difficult, dangerous, or repetitive for humans to perform alone. One of the key benefits of collaborative robots is their flexibility. They can be easily programmed and reprogrammed to perform a wide range of tasks. Another key benefit of collaborative robots is their ease of use. They are designed to be intuitive and easy to program, which means that even workers without specialised technical knowledge can quickly learn how to use them. By working together with humans, collaborative robots can help to reduce the risk of workplace injuries and improve overall productivity.

Industrial robotic technologies, as shown in Figure 11.6, provide several benefits, including:

1. **Increased productivity:** Robots can work 24/7 without any rest, which leads to increased productivity.
2. **Improved quality:** Robots can perform tasks with greater precision than human workers that leads to improvement in product quality.

3. **Increased safety:** Robots can perform dangerous and repetitive tasks that may be hazardous to human workers, which leads to increase in safety.
4. **Reduced labour costs:** Robots can perform tasks with greater speed and efficiency, reducing the needs for human workers and reducing labour costs.
5. **Increased flexibility:** Robots can be programmed to perform a wide range of tasks, which leads to increased flexibility in manufacturing and production processes.

11.4.10 Industrial machine learning technologies

Industrial machine learning technologies refer to the use of machine learning algorithms and techniques in industrial applications to improve manufacturing and production processes. These technologies enable machines to learn from data and improve their performance over time without being explicitly programmed [24].

There are several machine learning technologies, which are as follows:

Predictive maintenance: Predictive maintenance uses certain machine learning algorithms to predict equipment failure before they occur. By analysing data from sensors and other sources, predictive maintenance can identify patterns and anomalies that indicate a potential equipment failure. This technology can help reduce downtime, maintenance costs, and improve overall equipment efficiency.

Quality control: Quality control uses machine learning algorithms to detect defects and anomalies in manufactured products. By analysing data from sensors and cameras, quality control can identify defects and anomalies in real-time, allowing manufacturers to choose the right option quickly. This technology can improve product quality and reduce waste.

Process optimisation: Process optimisation uses machine learning techniques to identify inefficiencies in manufacturing and production processes. By analysing data from sensors and other sources, process optimisation can identify areas of improvement such as reducing energy consumption, improving cycle time, and reducing material waste. This technology can improve efficiency and reduce costs.

Supply chain optimisation: Supply chain optimisation technology uses machine learning techniques to optimise the supply chain by predicting demand, identifying the best suppliers, and improving logistics. By analysing data from multiple sources, supply chain optimisation can help the manufacturers improve their delivery times and reduce their inventory costs.

Autonomous control: Autonomous control uses machine learning techniques to enable machines to make decisions and take actions without human intervention. By analysing data from sensors and cameras, autonomous control can enable machines to make real-time decisions based on the current condition. This technology can improve efficiency and reduce labour costs.

Overall industrial machine learning technologies are essential for improving manufacturing and production processes. By enabling machines to learn from data and make real-time decisions, companies can reduce costs, improve efficiency, and remain competitive in the global marketplace. However industrial machine learning requires reliable data sources, skilled professionals to analyse and interpret data, and robust systems to implement machine learning algorithms.

11.4.11 Industrial Internet of Things technologies

Industrial Internet of Things (IIOT) is a network of connected devices, sensors, and machines that communicate with each other and collect data in real-time, enabling organisations to optimise their operations, reduce costs, and improve efficiency. The IIOT is often used in manufacturing, logistics, and other industries to improve productivity, reduce downtime, and increase profitability [20].

The IIOT consists of several components, including the following:

1. **Sensors and devices:** These are the physical components that are connected to the Internet and collect data from the environment. They can be temperature sensors, pressure sensors, vibration sensors, or other types of sensors that are used to measure various parameters.
2. **Data transmission and storage:** Once the data is collected by the sensors, it needs to be transmitted to a cloud based or on premises storage system. This data can be processed using various technologies such as cloud storage, edge computing, or fog computing.
3. **Data processing and analysis:** Once the data is stored, it needs to be processed and analysed to derive insights and actionable information. This can be done using advanced analytics, machine learning, or artificial intelligence technologies.
4. **Visualisation and reporting:** Once the insights and actionable information are derived, they need to be presented to the users in a way that is easy to understand. This can be done using dashboards, reports, or other types of visualisation.

11.5 CONCLUSION

11.5.1 Summary

The CPS is likely going in the form of AI, having improved security and privacy. Cyber security is expected to play a major role in the design and development of future engineering systems with new capabilities by exceeding the stages of autonomy, functionality usability reliability, and cyber security. Advances in CPS research can be extended with the aid of near collaborations between academic disciplines in computation, verbal exchange, management, and other engineering systems and computer technology disciplines.

11.5.2 Future trends in cyber-physical systems

Artificial intelligence is increasingly being integrated into CPS to enable intelligent decision-making automation. Machine learning algorithms can be trained on large data sets to predict future events and optimise the operation of physical systems. Predictive maintenance can be used to anticipate equipment failures and schedule maintenance before the failure occurs.

On the second hand, blockchain technology has the potential to enable secure end decentralised control of CPS. By using a distributed ledger to record transactions, it is possible to ensure the integrity and authenticity of data in CPS.

The next generation of cellular networks, 5G, promises to provide significantly higher data rates and lower latency than previous generations. This can enable real-time control of CPS over a wireless network enabling new applications and use cases.

As CPS became more intelligent and autonomous, it is important to ensure that they can work effectively with human operators. This requires new interfaces and communication protocols that enable effective collaboration between humans and machines.

REFERENCES

1. Sahra Sedigh, Ali Hurson, "Chapter one–Introduction and Preface", *Advances in Computers*, vol. 87, pp. 1–6, 2012.
2. Sahra Sedigh, Ali Hurson, "Cyber-Physical Systems: A Cyber-Physical System (CPS) is a system of collaborating computational elements controlling physical entities.", *Advances in Computers*, vol. 87, pp. 1–6, 2012.
3. Pardis Pishdad-Bozorgi, Xinghua Gao, Dennis R. Shelden, "Introduction to Cyber-Physical Systems in the Built Environment", *Construction 4.0*, pp. 23–41, February 2020.
4. Raj K Arora, "Cyber Physical Systems–Concept to rRality", 23 March, 2021. https://indiaai.gov.in/article/cyber-physical-systems-concept-to-reality

5. Amit Kumar Tyagi, N. Sreenath, "Cyber Physical Systems: Analyses, Challenges and Possible Solutions", *Internet of Things and Cyber-Physical Systems*, vol. 1, pp. 22–33, 2021.
6. A. Rahman, X. Gao, J. Xie, I. Alvarez-Fernandez, H. Haggi, W. Sun, "Challenges and Opportunities in Cyber-Physical Security of Highly DER-Penetrated Power Systems" *2022 IEEE Power & Energy Society General Meeting (PESGM)*, Denver, CO, USA, 2022, pp. 1–5, doi: 10.1109/PESGM48719.2022.9917144.
7. Sanjit A. Seshia, Shiyan Hu, Wenchao Li, Qi Zhu, "Design Automation of Cyber-Physical Systems: Challenges, Advances, and Opportunities" *IEEE Transactions on Computer-Aided Design of Integrated Circuits and Systems*, vol. 36, no, 9, September 2017.
8. Paulo Leitão, Armando Walter Colombo, Stamatis Karnouskos, "Industrial Automation Based on Cyber-Physical Systems Technologies: Prototype Implementations and Challenges", *Computers in Industry*, vol. 81, pp. 11–25, September 2016.
9. Salvador Cobos Guzman, Maria Dolores Cima Cabal, Francisco Machio Regidor, Lucia Alonso Virgos, "*Cyber-Physical System Architecture for Minimizing the Possibility of Producing Bad Products in a Manufacturing System*", 19 September, 2019.
10. Syed Hassan Ahmed, Gwanghyeon Kim, Dongkyun Kim, "*Cyber Physical System: Architecture, Applications and Research Challenges*", Conference: *Wireless Days (WD)*, IFIP, 2013.
11. Venkat N. Gudivada, Srini Ramaswamy, Seshadhri Srinivasan, "Data Management Issues in Cyber-Physical Systems", *Transportation Cyber-Physical Systems* (pp. 173–200), January 2018.
12. Muammer Eren Sahin, Lo'ai Tawalbeh, Fadi Muheidat, "The Security Concerns on Cyber-Physical Systems and Potential Risks Analysis Using Machine Learning", *Procedia Computer Science*, vol. 201, pp. 527–534, 2022.
13. Anthony Serino, Liang Cheng, "Real-Time Operating Systems for Cyber-Physical Systems: Current Status and Future Research" *International Conferences on Internet of Things (iThings) and IEEE Green Computing and Communications (GreenCom) and IEEE Cyber, Physical and Social Computing (CPSCom) and IEEE Smart Data (SmartData) and IEEE Congress on Cybermatics (Cybermatics)*, 28 December 2020.
14. Glenn A. Fink, Thomas W. Edgar, Theora R. Rice, Douglas G. MacDonald, Cary E. Crawford, "Overview of Security and Privacy in Cyber-Physical Systems", *Security and Privacy in Cyber-Physical Systems* (pp.1–23). 2017.
15. Andreas Burg, Anupam Chattopadhyay, Kwok-yan Lam, "Wireless Communication and Security Issues for Cyber–Physical Systems and the Internet-of-Things", *Proceedings of the IEEE*, vol. 106, no. 1, pp. 38–60, January 2018.
16. W. Bolton, *Programmable Logic Controllers*, Fourth Edition, Elsevier Newnes, ISBN-13: 978-0-7506-8112-4, 2006.

17. Christophe Kolski & Emmanuelle Le Strugeon, "A Review of Intelligent Human-Machine Interfaces in the Light of the ARCH Model" *International Journal of Human-Computer Interaction*, vol. 10, no. 3, pp. 193–231, 1998.

18. S. Kalaivani, M. Jagadeeswari, "PLC and SCADA Based Effective Boiler Automation System for Thermal Power Plant", *International Journal of Advanced Research in Computer Engineering & Technology*, vol. 6, no. 2, pp. 72–79, April 2018.

19. Zuzana Papulová, Andrea Gažová, Ľubomír Šufliarský, "Implementation of Automation Technologies of Industry 4.0 in Automotive Manufacturing Companies", *Procedia Computer Science*, vol. 200, pp. 1488–1497, 2022.

20. G. Lampropoulos, K. Siakas, T. Anastasiadis, "Internet of Things in the Context of Industry 4.0: An Overview.", *International Journal of Entrepreneurial Knowledge*, vol. 7, no. 1, pp. 4–19, 2019.

21. Belkacem Kada, Ahmed Alzubairi, Abdullah Tameem, "Industrial Communication Networks and the Future of Industrial Automation," *2019 Industrial & Systems Engineering Conference (ISEC)*, Jeddah, Saudi Arabia, pp. 1–5, 2019.

22. Theofanis P. Raptis, Andrea Passarella, Marco Conti, "Data Management in Industry 4.0: State of the Art and Open Challenges," *IEEE Access*, vol. 7, pp. 97052–97093, 2019.

23. M. Bahrin, M. Othman, N. Azli, M. Talib, "Industry 4.0: Review on Industrial Automation and Robotic", *Journal Teknologi*, pp. 137–143, 2016.

24. Thorsten Wuest, Daniel Weimer, Christopher Irgens, Klaus-Dieter Thoben, "Machine Learning in Manufacturing: Advantages, Challenges, and Applications", pp. 23–45, 24 June 2016.

Chapter 12

The impact of ubiquitous computing on heterogeneous next generation networks

Manivel Kandasamy[1], Raju Shanmugam[1],
Harshvi Adesara[1], Vishva Patel[1],
and Rajesh Kumar Dhanaraj[2]

[1]Unitedworld Institute of Technology, Karnavati University, Gandhinagar, Gujarat, India
[2]Symbiosis Institute of Computer Studies and Research (SICSR), Symbiosis International (Deemed University), Pune, India

12.1 INTRODUCTION

Heterogeneous next-generation networks (HetNets) are drastically changed by ubiquitous computing, which is the seamless integration of computing technology into everyday products and situations. A more connected eco-system is fostered by the abundance of smart devices and sensors in our environment, which leads to an increase in data traffic and calls for improved network efficiency. HetNets, which stand out for their adaptability and flexibility, are crucial for handling the various and changing connectivity requirements brought on by ubiquitous computing [1].

A unique computational paradigm known as ubiquitous computing will permeate all facets of modern life in the ensuing decades. From a technical standpoint, ubiquitous computing combines distributed systems, miniature hardware, and wireless networks and builds upon the technological advances made in these and other fields of engineering and research. The use of cutting-edge technologies, such as edge computing and artificial intelligence, is required in order to optimise network resources, guarantee real-time data processing, and provide individualised services as a result of this paradigm change. But this widespread connectedness also raises security and privacy issues, necessitating strong protection mechanisms within HetNets. HetNets open the door for cutting-edge services and applications that transform how we engage with the rapidly increasing digital world as they develop to meet the demands and opportunities provided by ubiquitous computing [2].

Additionally, careful network planning and deployment techniques within HetNets are necessary for the widespread adoption of ubiquitous computing. Ensuring seamless coverage and excellent performance across various environments becomes crucial as a variety of devices with different capabilities and connectivity requirements get interconnected. In order to

DOI: 10.1201/9781003362685-12

fulfill the needs for low latency services and real-time data processing, the integration of edge computing capabilities within HetNets becomes crucial, ushering in a new era of applications and services that take advantage of the power of the dispersed network. Service providers can exploit the data produced by ubiquitous devices to deliver personalised, context-aware services, increasing customer pleasure and opening up new business options.

Moreover, ubiquitous computing has a profoundly disruptive effect on heterogeneous next-generation networks, providing unmatched connection and opportunity while demanding careful planning, cutting-edge technology, and strong security measures within HetNets to fully realise its promise.

12.1.1 Evolution of ubiquitous computing

Since its conception, ubiquitous computing, sometimes known as "ubicomp," has seen a remarkable transformation. With its foundation in the idea of computers being effortlessly incorporated into daily life, ubiquitous computing has developed from theoretical principles to real-world applications that influence how people engage with modern technology [3].

The fundamental concepts of ubiquitous computing emerged in the latter half of the 20th century and were developed by scientists like Mark Weiser at Xerox PARC. In Weiser's vision, computer gadgets would blend into the background and become a widespread yet invisible component of the environment. A paradigm shift towards distributed and context-aware computing resulted from this concept's departure from the dominant paradigm of centralised computing [4].

With developments in areas like mobile computing, wearable technology, and sensor networks, ubiquitous computing has evolved inextricably. With their ability to provide real-time access to information, communication, and services, smartphones and wearable technology have emerged as leading examples of how computers may be integrated into daily life.

Furthermore, the development of artificial intelligence and machine learning has enabled ubiquitous computing systems to interpret the data they gather and adjust to user preferences and behaviour. The seamless integration of technology into several facets of daily life, from smart homes and healthcare to transportation and urban planning, is made possible in part by this adaptive intelligence.

The development of ubiquitous computing has taken a radical turn from theoretical ideas to real-world, adaptable systems that pervade many facets of contemporary life. This development is a reflection of the constant effort to make technology blend inconspicuously and flawlessly with our environment, improving the human experience in unimaginable ways [5].

12.2 KEY TERMINOLOGIES FOR UBIQUITOUS COMPUTING

There are several significant terms in the subject of ubiquitous computing that describe its fundamental ideas and technical components. Understanding these concepts is essential for comprehending the foundations and development of ubiquitous computing. The few key terminologies include context awareness, Internet of Things (IoT), pervasive computing is comparable to ubiquitous computing in that it emphasises technology's pervasive presence and seamless integration into a variety of situations, ubiquitous computing, smart environments is a technology that allows physical places to be monitored, analysed, and optimised for effectiveness, comfort, and user experience. Smart workplaces, smart cities, and smart homes are a few examples, sensor networks consists of networks of linked sensors that gather environmental data and transmit it for analysis and decision-making, adaptive intelligence is computer systems' potential way to learn from user interactions and gradually modify their behaviour and responses in order to better meet specific needs and preferences. Human-computer interaction (HCI) and calm technology is a design approach that tries to reduce pointless demands for users' attention from technology so they may continue to pay attention to their work and environment [6].

The phrase "Internet of Things" (IoT) emphasises the interconnection of real-world items that are equipped with sensors, software, and communication capabilities. This network is made possible by the ubiquitous interchange of data between devices. Similar to ubiquitous computing, pervasive computing emphasizes the idea that technology permeates every aspect of life and is integrated into a variety of settings and scenarios.

Wearable technology, which includes gadgets that people may wear to access computer capabilities while on the go, emerges as a concrete representation of ubiquitous computing. These gadgets, like smartwatches and glasses for augmented reality, are the perfect example of how computers and daily activities have merged. In turn, sensor networks serve as the framework for environmental data collecting, enabling real-time monitoring and well-informed decision-making.

Contextual computing and adaptive intelligence work together to create systems that can learn from user interactions and change their answers to suit different users' preferences. Within the scope of ubiquitous computing, this personalised approach promotes improved user experiences. In addition, the idea of "smart environments" dominates, where physical locations are outfitted with technology to enhance conditions for effectiveness, comfort, and general user wellbeing [7].

Ubiquitous computing has sparked the development of new paradigms in the field of human-computer interaction (HCI) that put a focus on usability, accessibility, and user pleasure. Additionally, ambient intelligence pushes the limits of ubiquitous computing by attempting to develop intelligent settings that automatically recognise and respond to human requirements.

Key terms in the field of ubiquitous computing come together to highlight its complexity. These concepts, which range from context awareness and the Internet of Things to wearable technology, sensor networks, and smart environments, all highlight how much the computer has permeated our daily lives and are influencing how humans engage with technology in the future [8].

12.2.1 Heterogeneous network architectures and implementations

In the world of contemporary networking, heterogeneous network designs and implementations create a dynamic environment. Contrary to conventional homogeneous networks, which are made up of standardized parts and protocols, heterogeneous networks include a variety of components that combine to generate flexible and effective communication infrastructures. To improve performance, coverage, and resource allocation, this strategy integrates many technologies, including cellular, Wi-Fi, and satellite networks.

The necessity to fulfil the increasing demands of data-hungry applications and an expanding number of connected devices is what propels heterogeneous network designs. Combining several network types builds on each technology's advantages while minimising its drawbacks and enhancing network capabilities. For instance, Wi-Fi networks give quick local access while cellular networks offer extensive coverage and mobility support. Smooth handoffs and load balancing may be accomplished by carefully installing these networks together, improving user experiences and effectively utilising resources.

The implementation of heterogeneous networks calls for complex issues with interoperability, administration, and optimisation. Intelligent routing algorithms, complex handoff mechanisms, and dynamic resource allocation strategies are all necessary for coordinating the operation of various network components. To provide smooth setup, monitoring, and integration of heterogeneous components, effective management and orchestration frameworks are also essential [8].

The introduction of fifth-generation (5G) cellular technology in recent years has sped up the development of heterogeneous networking. The flexibility, low latency, and high bandwidth features of 5G are well compatible with heterogeneous network theory. A key component of 5G, network slicing allows for the construction of virtualized and customised network segments, making it possible for numerous services with different needs to coexist in the same infrastructure [9].

The strategic strategy used to solve the complex requirements of modern connection is best shown by the diverse network designs and implementations. These networks have the ability to unleash improved performance, coverage, and user experiences by exploiting the synergies of many network technologies and using advanced management and orchestration methods, ushering in a new era of varied and effective communication paradigms [10].

12.2.2 Sensors and actuators and network models in ubiquitous computing

Sensors and actuators play a crucial role in ubiquitous computing, bridging the gap between the digital and physical worlds. Similar to the senses in a digital ecosystem, sensors observe and collect information about the world around them, recording elements like temperature, motion, and light. The system's decision-making mechanisms are then informed by these data streams, enabling context-awareness and adaptive behaviours. Actuators, on the other hand, serve as the system's effectors, converting digital orders into perceptible alterations in the external environment. These dynamic components, which might be motors, valves, or displays, cause reactions based on the evaluation of data from sensors. The cooperation of sensors and actuators enables ubiquitous computing systems to react independently and intelligently to changing circumstances [11].

Effective network models that enable smooth communication and interaction between sensors, actuators, and other computational entities are essential for ubiquitous computing. Ad hoc networks provide self-configuring connections that let devices directly communicate with one another without the need for a centralised infrastructure. They are a good option for mobile and dynamic settings in particular. Wireless sensor networks (WSNs) provide the foundation for applications requiring intensive monitoring since they collect data from dispersed sensors. Mobile ad hoc networks (MANETs), which support dynamic topologies, similarly enable direct peer-to-peer communication between mobile devices. In contrast, client–server architectures and cloud computing models support dispersed or centralised computation, allowing for effective data processing and interchange. By moving processing closer to data sources, lowering latency, and enhancing real-time capabilities, edge computing completes this landscape.

These network models operate together to coordinate data and communication flow, allowing sensors to gather information from the actual world, actuators to respond, and the entire ecosystem to work in unison. Thus, ubiquitous computing manifests as a symphony of connected components that perceive, analyse, and act – a seamless fusion of the digital and physical worlds.

12.2.3 Smart environments in heterogeneous future networks

Future heterogeneous networks' "smart environments" depict a paradigm in which intelligent technologies converge in real-world settings like homes, cities, and workplaces. By combining cutting-edge wireless technologies and communication protocols, heterogeneous future networks can take advantage of their wide range of possibilities. These smart environments allow for real-time monitoring, automation, and process optimisation by integrating edge computing, 5G and beyond connectivity, the Internet of Things (IoT), and powerful data analytics.

In the framework of smart environments in heterogeneous future networks, important elements and ideas include the following:

Internet of Things (IoT): IoT devices, which are networked objects and embedded sensors, are a key component of smart surroundings. To enable automation, monitoring, and control of numerous processes, these devices gather and communicate data. Different IoT device types with various communication needs can coexist and communicate effectively in a heterogeneous future network [12].

5G and beyond: Future heterogeneous networks frequently use the newest wireless communication technologies, including 5G and beyond. The various IoT devices and applications found in smart environments require networks that support high data rates, low latency, and widespread device connectivity.

Smart applications: Numerous applications can be found in smart settings, such as smart homes, smart cities, smart factories, and more. These applications could pertain to, among other things, infrastructure management, environmental monitoring, healthcare, and transportation.

Security and privacy: It is crucial to provide security and privacy in smart environments due to the development of linked devices and data sharing. Strong security measures must be implemented in heterogeneous networks to safeguard critical information and stop illegal access [13].

Moreover, in order to deliver smooth and effective connectivity, these smart settings make use of the capabilities of heterogeneous future networks, which are sophisticated and diverse communication infrastructures that combine a variety of different wireless technologies, protocols, and devices [14].

Applications such as energy management, healthcare, and urban planning are made possible by this revolutionary synergy, while strong standards guarantee security and privacy. The promise of these smart settings is further amplified by device compatibility across several wireless standards, improving overall productivity, comfort, and sustainability in our rapidly changing technological environment [15].

12.3 APPLICATIONS OF HETNETS

Numerous cutting-edge applications can now be developed thanks to the impact of ubiquitous computing on heterogeneous next-generation networks (HetNets), which takes advantage of the seamless integration of smart devices and networked environments. HetNets and ubiquitous computing enable telemedicine and remote patient monitoring, transforming patient care, and lowering healthcare expenses. Businesses gain from increased productivity by monitoring and controlling machinery and processes in real-time [16].

12.3.1 Smart cities

Smart cities are made possible by ubiquitous computing, where different IoT devices and sensors are linked together by HetNets.

These networked systems can boost public safety through smart surveillance, improve waste management, and optimise traffic flow in addition to giving citizens real-time information on the weather, events, and public transportation.

A paradigm for urban development known as "smart cities" uses technology, data, and connection to improve livability for citizens while streamlining the use of resources and city services. Urban planning is being transformed by ubiquitous computing and next-generation networks, which are essential to achieving the goal of smart cities.

The seamless integration of computer devices and sensors throughout the urban environment, known as ubiquitous computing, makes it possible to gather and analyse massive volumes of real-time data. It is possible to use this data to learn more about urban dynamics, including traffic patterns, energy use, trash management, and air quality. These perceptions serve as the cornerstone for making well-informed choices in urban planning. Moreover, high-speed broadband, 5G, and other next-generation networks enable the effective transfer of data between various devices and sensors. This network architecture enables real-time communication between gadgets, data centres, and city management systems while supporting the connection needed for smart city applications. These networks enable applications like intelligent traffic management, remote healthcare services, and real-time environmental monitoring thanks to their low latency and high bandwidth.

Urban planning as shown in Figure 12.1 benefits from ubiquitous computing and next-generation networks include the following:

Data-driven decision making: Urban trends and problems may be understood by analysing the real-time data gathered by ubiquitous sensors. This data-driven strategy enables planners to choose wisely when it comes to resource allocation, infrastructure development, and service enhancement.

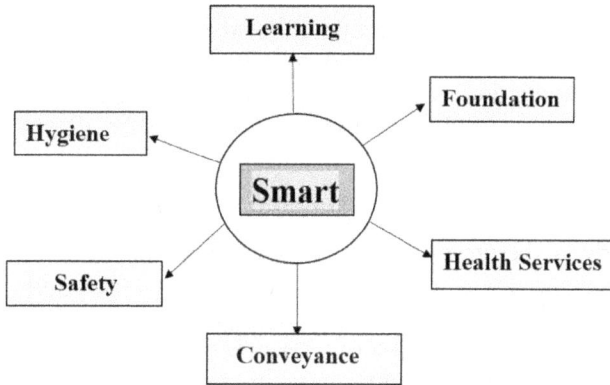

Figure 12.1   Revolutionising urban planning.

Effective resource management: By examining data on electricity, water, and trash generation, smart cities can maximise resource utilization. This knowledge may direct efforts to cut waste, save resources, and cut expenses all around.

Disaster management and resilience are two areas where smart city technology can help. During natural catastrophes, real-time data analysis can help with early warning systems and evacuation planning.

However, there are difficulties with integrating next-generation networks and ubiquitous computing into urban design. To promote fair and sustainable smart city development, it is important to take into account a number of issues, including privacy concerns, data security, network dependability, the digital divide, and the possibility for technology-driven exclusions.

In conclusion, ubiquitous computing and next-generation networks are allowing the development of smart cities that are data-driven, effective, sustainable, and sensitive to the demands of their citizens. This is revolutionising urban planning. These ideas will continue to influence urban life in the future as technology develops.

12.3.2 Healthcare

HetNets and ubiquitous computing enable telemedicine and remote patient monitoring in the healthcare industry. Healthcare practitioners can monitor patients' health conditions in real-time using wearable technology and health sensors, which can continually collect important health data, securely communicate it over the network, and improve patient outcomes and lower healthcare costs.

12.3.3 Industrial IoT

HetNets are used by ubiquitous computing applications in industrial environments to offer real-time machine and process monitoring and control. This boosts production and cost-effectiveness for industries by reducing downtime, improving efficiency, and enabling predictive maintenance.

12.3.4 Improved quality of service (QoS)

A key objective of the integration of ubiquitous computing with next-generation networks is to increase quality of service (QoS). The capacity of a system to deliver dependable, effective, and predictable performance, guaranteeing that users obtain the anticipated level of service, is referred to as QoS. Advanced networks and ubiquitous computing have the potential to greatly improve QoS in a number of ways. Improved connectivity and coverage: 5G and other next-generation networks provide greater data speeds, lower latency, and better coverage. This makes it possible for a variety of devices to connect seamlessly in scenarios involving ubiquitous computing. Even in isolated places or heavily crowded cities, users can count on constant and dependable connections.

High bandwidth: In ubiquitous computing, a lot of data is transferred between linked devices. Higher bandwidth is offered by next-generation networks, enabling the efficient transport of huge files, multimedia content, and data-intensive applications.

Resource allocation and optimisation: Systems for ubiquitous computing produce a lot of data flow. Next-generation networks can optimise network resources to avoid congestion and sustain QoS, particularly during periods of high demand.

Edge computing: When ubiquitous computing and edge computing are combined, data processing may be done closer to the data source, lowering latency and boosting QoS for real-time applications that need quick answers.

Dynamic network configurations: Mobile devices and shifting network topologies are frequent components of ubiquitous computing. As users move between various service regions, next-generation networks can react dynamically to these changes, enabling smooth handovers and maintaining a constant QoS.

Despite these advantages, it is important to carefully handle issues like security flaws, network congestion, interoperability, and the possibility of digital inequities in order to make sure that QoS enhancements are fair and long-lasting. In a variety of industries, such as entertainment, healthcare, transportation, and industrial automation, the integration of ubiquitous computing with next-generation networks has the potential to greatly improve the quality of service. The emphasis on QoS improvement will be a driving factor in forming the future of networked and technologically sophisticated settings as these technologies continue to develop.

12.3.5 Static and dynamic spectrum allocation

In the context of ubiquitous computing and heterogeneous next-generation networks, where effective use of the radio frequency spectrum is essential, static and dynamic spectrum allocation are key topics. These methods tackle the problem of meeting the various communication requirements of a variety of devices while preserving optimal performance and reducing interference.

Frequency allocation, commonly referred to as static spectrum allocation, is assigning particular frequency bands to different communication services. Usually, licencing, standards, and regulatory decisions provide the basis for this set distribution. With this method, distinct frequency bands that are less prone to interact with one another are allotted to various services or technologies. Although static allocation offers consistency and predictability, it may lead to underuse of the spectrum, particularly if some frequency bands are crowded while others are free.

Moreover, the usage of the spectrum is made flexible and opportunistic by dynamic spectrum allocation, also known as dynamic spectrum sharing or spectrum access. This strategy is especially useful in situations when there are scarce spectrum resources and demand fluctuates according to place and time. Utilising cutting-edge technology like cognitive radios and software-defined radios, dynamic allocation senses the spectrum environment and assigns frequencies dynamically based on current circumstances. As a result, devices that require certain frequencies may access those that are underutilised or infrequently used, enabling more effective spectrum utilisation

Advantages of dynamic spectrum allocation

a. Efficiency: By enabling devices to access unused frequencies, dynamic spectrum allocation optimises spectrum consumption, improving overall network efficiency.
b. Flexibility: Equipment can adjust to shifting spectrum circumstances, lowering interference and enhancing the clarity of communications.
c. Spectrum sharing: By using dynamic spectrum sharing, many technologies and services may coexist more peacefully.
d. Situational adaptation: In reaction to shifting environmental variables, devices can dynamically move to less crowded frequencies.

Challenges of dynamic spectrum allocation

a. Interference: In order to stop detrimental interference between devices that opportunistically use the spectrum, dynamic allocation requires sophisticated interference control solutions.

b. Broadcast sensing accuracy: For efficient dynamic allocation, accurate spectrum sensing is necessary. Inaccurate sensing may cause licenced users to be interfered with.

c. Coordination: To avoid interference and provide equal spectrum access for competing devices, effective coordination techniques are needed.

So, both static and dynamic spectrum allocation schemes have their benefits and drawbacks. The decision between these strategies depends on a number of variables, including the availability of spectrum, regulatory issues, technological limitations, and the desired degree of flexibility and efficiency in spectrum utilisation.

12.3.6 Augmented reality (AR) virtual reality (VR)

HetNets and ubiquitous computing work together to support the growth of AR and VR applications. High-speed and low-latency networks enable seamless delivery of AR and VR experiences, increasing collaboration, training, and entertainment across multiple industries.

12.3.7 Transportation and logistics

HetNets and ubiquitous computing together improve logistics and transportation operations. Utilising real-time data from several sensors and devices, smart transportation solutions optimise traffic flow, enhance public transportation services, and facilitate effective fleet management.

12.3.8 Retail and marketing

Retailers may provide customers with customised shopping experiences thanks to ubiquitous computing. Retailers can make specialised recommendations based on customer preferences and behaviours, enhance store layouts, and offer targeted discounts by utilising data from connected devices.

12.3.9 Environmental monitoring

HetNets and pervasive computing enable thorough environmental monitoring. Various sensors and gadgets can gather information on water levels, weather, and air quality, facilitating better environmental management and decision-making.

As technology develops, the coming together of ubiquitous computing and HetNets is expected to spark a surge of innovation that will enable people, organisations, and societies to use data-driven intelligence to promote society's goals and build a more interconnected and sustainable future. The development of sophisticated artificial intelligence (AI) systems that process and analyse enormous amounts of data in real-time to produce insightful results and automate difficult activities is also made possible by the combination of HetNets and ubiquitous computing.

Additionally, ubiquitous computing's influence on HetNets goes beyond specific applications, supporting a more comprehensive transformation of society and industry. By enhancing infrastructure, energy efficiency, and citizen engagement, smart cities transform urban living. Healthcare systems become more patient-centric and cost-effective, with telemedicine and remote monitoring enhancing accessibility to quality care. The productivity of industrial sectors rises, downtime is decreased, and resource allocation and production methods are optimised. Smart device and sensor integration into daily life generates enormous networks of data that may be used to inform data-driven decision-making across a variety of disciplines. Predictive analytics, for example, can be performed on data gathered from connected devices to help businesses and organisations predict trends and customer behaviour, improve operations, and provide customised services.

The development of sophisticated artificial intelligence (AI) systems that process and analyse enormous amounts of data in real-time to produce insightful results and automate difficult activities is also made possible by the combination of HetNets and ubiquitous computing. These applications go beyond established industries and into new ones like smart agriculture, where HetNets enable precision farming methods by giving real-time information on crop health, weather patterns, and soil conditions. As technology develops, the coming together of ubiquitous computing and HetNets is expected to spark a surge of innovation that will enable people, organisations, and societies to use data-driven intelligence to promote society's goals and build a more interconnected and sustainable future.

Overall, ubiquitous computing and HetNets' synergy pushes us toward a more connected, intelligent, and sustainable future where technology effortlessly enhances societal challenges and quality of life. These applications' combined influence will transform global industries, economies, and communities as they continue to develop and converge.

12.3.9.1 Integration of 5G with previous generations

An important development in the evolution of global connection has been made with the combination of 5G with earlier generations of wireless communication technology. This seamless fusion of the strengths of current wireless networks, particularly 4G LTE, with the innovative features and capabilities inherent in the fifth generation of wireless technology is frequently referred to as 5G progression.

A number of key ideas serve as the foundation for this integration. The cornerstone of a smooth transition between 4G to 5G networks for consumers and devices is backward compatibility. The fundamental 4G infrastructure is initially used by 5G networks in a non-standalone (NSA) mode for certain purposes, accelerating the spread of 5G services and maximizing earlier investments.

Additionally, the merging of various generations enables connectivity fluidity, enabling devices to easily move between network standards and coverage zones. This promotes a dependable and consistent user experience. This integration enables network operators to strategically inject 5G services into particular regions or industries while maximising current infrastructure thanks to its economical rollout and phased tactics.

In conclusion, combining 5G with earlier wireless generations is a smart synergy that makes use of both the strength of cutting-edge 5G technology and the solid basis of current networks. This strategy not only speeds up the switch to 5G but also improves the network's overall performance, creating opportunities for cutting-edge services and applications that are specifically suited to the various demands of people, businesses, and societies on a worldwide level.

12.3.9.2 Integration of Wi-Fi and IoT with various connections

The proliferation of networked devices and intelligent technologies is accelerated by the combination of Wi-Fi and the Internet of Things (IoT) with a wide range of connections. This integration creates an environment that supports the many and changing requirements of IoT applications by smoothly fusing the capabilities of Wi-Fi networks with a spectrum of communication possibilities.

As a widespread and fast wireless communication option, Wi-Fi is the key component of local area networks (LANs) in homes, places of business, and public areas. When Wi-Fi and IoT are combined, a strong ecosystem develops that takes advantage of Wi-Fi's well-established infrastructure to foster the growth of smart devices. This symbiosis offers a wide range of concrete advantages, such as ubiquity coverage that seamlessly connects a large number of Internet of Things devices and supports applications in a variety of contexts.

Wi-Fi is particularly suited for IoT applications like video streaming and large-scale data transfers because of its inherent high transmission rates, which are used to enable real-time communication and data-intensive operations. Wi-Fi's user-friendly setup and configuration procedures also make it possible for IoT devices to be seamlessly integrated, which encourages widespread consumer adoption.

Through numerous connections that are adapted to certain IoT use cases, the simultaneous integration of Wi-Fi and IoT expands its reach. The advantages of Wi-Fi are combined with the unique qualities of other communication technologies in this fusion. Beyond the limitations of conventional Wi-Fi networks, cellular networks, from 3G and 4G to the cutting-edge capabilities of 5G, offer extensive coverage for IoT deployments, supporting applications like asset monitoring and smart cities. Bluetooth and Bluetooth

Low Energy (BLE) network integration becomes beneficial in situations when energy efficiency and close device proximity are required. Wearables, medical equipment, and smart home automation are all supported by these short-range connections that tidily link nearby items.

Zigbee and Z-Wave, low-power mesh network technologies, are smoothly incorporated into this integration and serve industries including home automation and industrial control systems. These connections enable fluid communication between numerous IoT devices in small locations. To bring the Internet of Things (IoT) into tough and distant areas, long-range, low-power solutions like LoRaWAN and NB-IoT are effortlessly combined. Applications like smart meters, environmental monitoring, and asset management are all supported by this connection.

In summary, by combining Wi-Fi and IoT with various connections, a harmonious orchestration is created that best utilises the advantages of each communication technology. In addition to enabling IoT devices to effortlessly navigate different contexts, this collaborative fabric also prepares the way for a wide range of applications and sectors, supporting the development and maturation of the interconnected world that defines the IoT era.

12.3.9.3 Challenges in HetNets

Heterogeneous networks (HetNets), which include several cell types and wireless technologies, provide the promise of improved connectivity, coverage, and capacity. However, this integration also brings with it a number of complex issues that demand careful thought. As the presence of macrocells, microcells, and picocells can result in overlapping coverage areas and potential signal interference, managing interference is one of the biggest issues. To maintain ideal spectrum utilisation and network performance, effective interference mitigation solutions are crucial.

HetNets must also overcome problems with seamless mobility and handover management. It is challenging to provide seamless switching between several cell types while maintaining uninterrupted service, especially in conditions when user mobility is considerable. Furthermore, the importance of having a strong backhaul infrastructure cannot be overstated. A high-capacity, low-latency backhaul network is necessary for efficient data transfer between various cell types and the core network, but in some regions, establishing one can be difficult.

Effective resource management and allocation become crucial issues in HetNets. In order to respond in real-time to changing traffic demands and quality standards, sophisticated algorithms are needed to dynamically distribute resources like frequency bands and time slots across numerous cells and technologies. A sustainable approach that strikes a balance between performance and energy consumption is required as a result of the integration of multiple cell types with different power requirements.

HetNet implementation and planning require close attention to every last aspect. A thorough and systematic strategy is needed to determine the best cell sites, transmission powers, and antenna configurations for achieving seamless coverage and capacity across various terrains and circumstances. In order to reduce interference and achieve optimal network performance, it is essential to provide harmonious coordination and collaboration between various cell types and base stations.

Effective spectrum management is essential to avoid congestion and guarantee fair access for all users as HetNets continue to develop. HetNets' complexity and prospective costs should also be carefully considered, as the advantages of better performance must be evaluated against the necessary investments. A seamless and coherent network architecture is made possible by standardisation and compatibility among multiple technologies and providers.

Although HetNets have the potential to completely transform wireless communication, they also present a variety of difficulties. To successfully overcome these obstacles, a comprehensive strategy that integrates cutting-edge technologies, teamwork in research, and creative solutions is required. This will allow HetNets to reach their full potential and open the door to a connected and effective future.

12.4 ENHANCED DATA COLLECTION AND ANALYSIS OF UBIQUITOUS COMPUTING

The cornerstone of ubiquitous computing, a concept that smoothly integrates computational capabilities into daily living, is improved data collecting and processing. In this setting, a complex network of embedded sensors and gadgets continuously collects data, creating a pervasive stream of information. This data, which comes from various sources like motion sensors, cameras, and GPS trackers, is distinguished by its pervasiveness and real-time nature. The difficulty comes from handling the large amount of data as well as from drawing insightful conclusions from it.

This calls for advanced methods of data processing, analysis, and storage that frequently deal with big data problems. Contextual awareness is important, and data collection includes not just the what of user interactions but also the when, where, and how. This data is converted into useful insight using machine learning and AI, enabling tailored interactions, predictive models, and even real-time decision-making. However, privacy and security problems accompany the potential for improved data analysis in ubiquitous computing, calling for strong safeguards for private data. Effective data gathering and insightful analysis are the cornerstones of this technology's capacity to improve user experiences and streamline operations as it finds applications in smart homes, healthcare, transportation, and beyond.

12.4.1 Heterogeneous environments in ubiquitous computing

In ubiquitous computing, heterogeneous environments refer to the coexistence of many devices, technologies, and communication protocols inside a single computational environment. The goal of ubiquitous computing is to create a linked, intelligent ecosystem by seamlessly integrating computational capabilities into commonplace objects, settings, and activities. As a natural outcome of this situation, heterogeneous settings with a variety of traits and difficulties emerge.

Device diversity: A wide range of gadgets, including smartphones, wearables, sensors, actuators, appliances, and more, are included in heterogeneous environments. A difficult integration scenario might result from the variety of capabilities, form forms, and communication interfaces that these devices can have.

Communication protocols: Devices utilising various communication protocols, including Bluetooth, Wi-Fi, Zigbee, and NFC, are frequently used in ubiquitous computing. These protocols allow for data exchange and communication between devices, but due to their heterogeneity, effective interoperability depends on careful orchestration.

Data integration: The format, frequency, and purpose of the data generated by various devices in a heterogeneous environment can vary. For this diverse information to yield useful insights, efficient data integration and aggregation procedures are required.

Application development: It's important to take into account the strengths and weaknesses of various devices and technologies when creating apps for heterogeneous environments. The development of dependable and adaptable apps can be facilitated by cross-platform development tools and frameworks.

User experience: Despite their diversity, the smooth interaction between devices enhances the user experience. It is important to take into account the various capabilities and characteristics of devices when designing intuitive interfaces and interactions.

As a result of the wide range of devices, technologies, and communication channels that come together to form a linked ecosystem, heterogeneous environments are a defining feature of ubiquitous computing. The ideal of seamless integration and intelligent interactions, where computing capabilities effortlessly enhance numerous aspects of daily life, requires effective management of this heterogeneity.

12.4.2 Integrated connectivity and user experience

Central themes that emerge at the intersection of ubiquitous computing and next-generation networks are integrated connectivity and user experience.

They work in unison to produce a potent synergy that transforms how people engage with technology and their surroundings.

A future of ubiquitous computing is one in which technology is widespread and essentially invisible, blending effortlessly into daily life. In order for these various devices, sensors, and objects to interact and work together in real time, integrated connectivity is crucial to this vision. Wi-Fi, cellular networks, Bluetooth, and cutting-edge technologies like 5G and beyond are just a few examples of the extensive range of wireless technologies that are included in this connectivity. Because these technologies are integrated, connectivity is constant and seamless, enabling users to switch between devices and places without losing their digital identity.

An unmatched user experience sits at the core of this integration. The goal of ubiquitous computing, which is supported by integrated connectivity, is to improve how people interact with technology and their surroundings. These characteristics define this transformation, seamless interaction, fluid mobility, context-awareness, enhanced productivity and immersive experiences.

Integrated connection and user experience are even more revolutionary in the context of next-generation networks, such as 5G. The potential of ubiquitous computing is further amplified by 5G's high-speed, low-latency, and huge device connectivity capabilities, which allow for unparalleled real-time interactions and immersive experiences. With the capabilities of next-generation networks and the combination of connectivity and user experience in ubiquitous computing, a new age of frictionless interactions, personalised services, and improved daily life is introduced. The potential for building a future in which technology becomes an essential and empowering aspect of the human experience lies in this revolutionary convergence.

12.4.3 Role of edge computing

By decentralising computation and data processing closer to the data source, edge computing plays a crucial role in updating and optimising the digital landscape. This paradigm shift has a number of advantages that have a big impact on different areas of technology and business. Edge computing lowers latency by processing data at or near the edge of the network, enabling real-time responsiveness for applications that call for rapid decision-making, such autonomous vehicles and industrial automation. By sending only pertinent and condensed data to central servers, it also lessens the strain on network traffic. This optimises network resources and reduces congestion.

Since sensitive data can be processed locally, edge computing also plays a role in improving privacy and security by reducing the dangers involved in sending sensitive information over a network. This is especially important in industries like healthcare, where maintaining data privacy is of utmost importance. Edge computing also gives devices the ability to keep working even when they are not linked to the main network, guaranteeing continuous operation in circumstances with sporadic connectivity or remote locations.

Edge computing is a key component of the Internet of Things (IoT) industry. Before transferring critical data to the cloud, it enables IoT devices and sensors to process and filter data at the edge. As a result, network resources are not only conserved, but real-time analytics and insights are also made possible at the site of data collection, which is necessary for applications such as predictive maintenance as well as data-driven decision-making.

Additionally, edge computing improves interactivity and user experiences in interactive technologies like augmented reality (AR) and virtual reality (VR) by reducing latency. Edge computing's contribution to the responsive and effective digital ecosystem, where data processing as well as analysis take place seamlessly and intelligently at the edge of the network, is growing as the digital landscape changes.

12.4.4 Wireless devices for network optimisation

Wireless devices play a critical role in enhancing user experiences and maximising network performance in the context of ubiquitous computing and the transition to next-generation networks. These gadgets work together as a varied group, each adding something special to the complex web of network optimisation. The environment is filled with sensors and actuators that provide real-time data on things like user behaviours and environmental conditions. This information supports contextual awareness and enables dynamic network parameter adjustments for improved effectiveness and responsiveness.

Personalised network resource allocation and location-based services are made possible by personal devices, like as smartphones and wearables, which act as user-centric data collectors and provide insights into mobility habits and preferences. These gadgets can also serve as middlemen in decentralised communication, enhancing connectivity in difficult places. As they soar through the air, UAVs and drones provide temporary network assistance, quick reaction capabilities, and environmental data collection – all essential for maximising network coverage and performance. Internet of Things (IoT) technology, on the other hand, takes the form of intelligent gadgets that orchestrate energy-efficient data transmission and enable communication between connected cars or smart appliances. As localised processing powerhouses, fog and edge nodes minimise latency by doing real-time analysis near to the data sources. This hierarchical architecture supports time-sensitive applications and improves network responsiveness. Smartwatches and other wearable mesh networks enable direct, dependable device-to-device communication, which lowers congestion and improves network performance.

In the midst of this convergence, beamforming antennas excel by boosting coverage, reducing interference, and optimising signal transmission. This collection of wireless gadgets works as a whole to create a symphony of optimisation, reshaping networks into fluidly interconnected, flexible, and user-centric domains that are ready to handle the demands of the era of ubiquitous computing and the advent of next-generation networks.

12.5 SECURITY AND PRIVACY CONCERNS IN UBIQUITOUS NETWORKS

In ubiquitous networks, also known as ubiquitous computing or the Internet of Things (IoT), linked objects may smoothly exchange data and communicate with one another to improve a variety of elements of daily life. The enormous volume of data created, communicated, and retained by these networks raises serious security and privacy problems in addition to their host of other advantages.

The vast volume of data shared between these linked devices poses a serious threat to the privacy of personal data. This information might include private routines and sensitive location information, making it vulnerable to misuse and unauthorised access. Additionally, the multiplied attack surface brought on by the wide range of networked devices increases the risk of data breaches and highlights network flaws. These hazards are exacerbated by inadequate encryption techniques and ineffective security procedures, which allow attackers to intercept and alter data transfers.

Inconsistencies in protective measures are introduced by the absence of standardised security standards and differences in security implementations across devices. Additionally, attacks like malware injection and remote vulnerabilities may be successful against IoT devices with limited resources. Users frequently have little control over how data is collected and used, creating opportunities for privacy abuses. The network's extensive interconnections imply that a security flaw in one device may set off a series of events that could compromise the entire system. In order to ensure people's security and privacy within the ubiquitous network paradigm, addressing these challenges calls for a comprehensive approach that includes strong encryption, strict privacy-oriented design principles, frequent security updates, user education, and transparent data usage policies.

12.5.1 Scalability of complex systems

In the field of ubiquitous computing, scalability is a crucial factor to be taken into account when designing and implementing complex systems, especially when using heterogeneous next-generation networks. The goal of ubiquitous computing is to create a world where linked devices communicate with one another invisibly to improve user experiences. The need for scalable systems is growing in importance as the variety and number of devices increase. This problem is made more difficult by heterogeneous next-generation networks, which diversify the ecosystem's technologies, communication protocols, and device capabilities.

It takes a combination of important tactics to achieve scalability. First, modularity and adaptability must be considered when designing the system architecture. Because modular components can be scaled individually, resources can be allocated effectively according to demand. Additionally, the system needs to have dynamic resource allocation components that evenly share computing and communication resources among devices. The system will be able to gracefully handle changes in network traffic and device engagement because of its versatility.

Data management is another essential scaling component. Effective data processing, storage, and retrieval procedures are essential in the ubiquitous environment as data comes from various sources. In order to improve data processing capabilities, minimise bottlenecks, and guarantee timely access to information, scalable databases, distributed file systems, and caching techniques can be used.

The necessity to solve potential synchronisation and communication difficulties is inherent in the scaling difficulty. Maintaining coordinated and coherent interactions gets harder as the system gets bigger. To reduce dependencies and streamline interactions, scalable systems frequently use event-driven designs and asynchronous communication paradigms.

It takes a comprehensive approach to scale complex systems running on heterogeneous next-generation networks for ubiquitous computing. These systems can successfully navigate the challenges posed by an ever-expanding ecosystem of interconnected devices and diverse network landscapes thanks to strategic architecture design, dynamic resource allocation, effective data management, thoughtful synchronisation mechanisms, and the integration of cutting-edge technologies.

12.5.2 Management of a vast number of devices

Data storage and blockchain-based authentication improve security and data integrity in this complex ecosystem. The administration of a wide variety of ubiquitous computing-related devices has grown to be a complex task, especially given how smoothly these devices are woven into the architecture of heterogeneous next-generation networks (HNGNs). Effective management methods are essential to enable efficient operations and the best user experiences in this dynamic environment, which is defined by a variety of device kinds, network technologies, and user requirements.

A world of ubiquitous computing is one in which a variety of devices, from wearable technology to IoT sensors to conventional computers and smartphones, work together to deliver personalised and context-aware services. A strong management system that can handle device discovery, authentication, provisioning, monitoring, and maintenance is required due to the sheer number and variety of these devices. HNGNs, which combine 5G and other technologies, increase complexity because they provide different levels of bandwidth, latency, and coverage.

A multifaceted approach to device management is essential to overcoming these obstacles. Real-time insights into device behaviour can be obtained through centralised management systems, supported by artificial intelligence and machine learning, allowing for resource allocation and preventative maintenance. Decentralised solutions, like as distributed

On heterogeneous next-generation networks, managing a large number of devices in the context of ubiquitous computing requires a comprehensive strategy that combines centralised and decentralised management methods. In order to provide effective device orchestration, excellent resource usage, and a seamless user experience in this complex and changing environment, a resilient management framework can be built by combining the capabilities of AI, blockchain, SDN, NFV, and edge computing.

12.5.3 Network operations: Energy efficiency and sustainability

Energy efficiency and sustainability are essential for network operations in the context of ubiquitous computing on heterogeneous next-generation networks (HNGNs). The amount of energy consumed by these complex networks increases along with the development of devices and data traffic. Strategies that reduce environmental impact and maximise energy efficiency are essential for ensuring long-term viability. HNGNs, which are characterised by a variety of technologies and different energy needs, need for creative solutions to balance the rising energy needs of infrastructure and devices.

Energy efficiency improvements involve a variety of methods. Devices with adaptive sleep modes and dynamic resource allocation can reduce energy waste, and networks with sophisticated traffic management and load balancing can operate more efficiently.

Beyond energy efficiency, sustainability demands a comprehensive perspective of the full network lifecycle. This includes effective network deployment to reduce resource consumption, responsible disposal or recycling of out-of-date equipment, and responsible procurement of resources for network gear.

The management of network operations within ubiquitous computing on HNGNs depends on a dual emphasis on sustainability and energy efficiency. These networks may efficiently balance their energy demands while reducing their environmental impact by utilising adaptive resource management, renewable energy integration, and responsible network lifecycle practices.

Such tactics are essential to the HNGN paradigm's ethical and long-term survival in a time when digital connectivity must coexist with environmental concerns.

12.5.4 Network architecture innovations

To accommodate the requirements of ubiquitous computing and diverse next-generation networks, network architectural advances are essential. With the large range of devices and applications that these settings support, these technologies hope to guarantee smooth communication, effective resource utilisation, and optimised performance. Here are some significant advancements in network architecture under this situation:

a. Network slicing: Network slicing, a term used in 5G and later networks, refers to the division of a network into a number of virtualised networks, each suited to a different set of needs. This makes it possible for a variety of services with different performance requirements to coexist, such as enhanced mobile broadband (eMBB) for high-speed data services and ultra-reliable low-latency communication (URLLC) for essential applications.

b. Edge computing: Bringing computer power closer to data sources and end users is known as edge computing. This lowers latency, improves real-time processing capabilities, and enables applications like IoT devices, augmented reality, and crucial industrial processes that need for quick replies.

c. Fog computing: Fog computing extends computing capabilities to the edge of the network and integrates them with cloud resources, similar to edge computing. This strategy is especially pertinent in situations when localised processing is required due to latency, bandwidth restrictions, and data privacy concerns.

d. HetNet architecture: HetNet topologies combine several access technologies, including Wi-Fi, small cells, and cellular networks, to offer seamless coverage and capacity in a variety of urban and rural contexts.

e. Dynamic spectrum sharing: Multiple technologies, including 4G and 5G, may effectively share the same frequency bands thanks to dynamic spectrum sharing techniques, which also offer backward compatibility.

f. Network function virtualisation: Routing, firewalling, and load balancing are examples of classic network services that are virtualised and operate on affordable hardware as part of NFV. This increases network service deployment's flexibility, scalability, and affordability.

The various and dynamic characteristics of ubiquitous computing and heterogeneous next-generation networks provide difficulties that are collectively addressed by these network architectural improvements. These technological advancements open the door for the realisation of cutting-edge applications and services across several fields by offering adaptive, effective, and dependable connection solutions.

12.5.5 Network optimisation and predictive analysis

It is possible to achieve seamless connection and higher performance in the setting of ubiquitous computing and heterogeneous next-generation networks thanks to the synergy between network optimisation and predictive analysis. The term "network optimisation" refers to a group of techniques for fine-tuning energy efficiency, load distribution, quality of service management, and network resources. It makes sure that the different devices' and applications' varied communication needs are satisfied while maximising resource use and maintaining a high level of service. In parallel, predictive analysis makes use of both historical and present data to foresee prospective problems as well as future trends and behaviours. By predicting peak use periods, seeing security risks, and foreseeing equipment problems, this predictive strategy helps with proactive network management.

Network operators can anticipate problems, distribute resources wisely, and offer a smooth user experience by successfully combining network optimisation with predictive analysis. This cooperative strategy is essential for navigating the difficulties of ubiquitous computing and heterogeneous networks, which eventually leads to improved user happiness and optimised network performance.

12.5.6 Dynamic spectrum management for efficient resource allocation

In the dynamic environment of ubiquitous computing and heterogeneous next-generation networks, dynamic spectrum management is essential for optimising resource allocation. Effective spectrum utilisation becomes crucial as different devices and applications proliferate. Based on current demand and network conditions, dynamic spectrum management provides adaptive and real-time frequency band allocation. By letting various technologies and services to live peacefully and decreasing spectrum waste and congestion, this strategy improves resource efficiency.

Additionally, it supports the various communication needs of IoT devices, mission-critical applications, and high-speed data services, guaranteeing that each obtains the required bandwidth while preventing interference. Dynamic spectrum management gives network operators the ability to adapt to shifting traffic patterns, provide priority to vital services, and proactively handle network congestion. This coordination of spectrum resources creates a robust environment, optimises connection, reduces latency, and meets the changing requirements of ubiquitous computing and heterogeneous next-generation networks.

12.5.7 Regulatory and policy implications

To ensure fair access, data privacy, security, and effective network management, the deployment of ubiquitous computing and heterogeneous next-generation networks raises important legislative and policy considerations. These ramifications cover a range of technological, social, and user rights topics:

a. Spectrum management and allocation: Effective spectrum management rules are required due to the rising demand for spectrum resources. The needs of diverse services and technology must be balanced by regulators while unwanted intervention is avoided. To maximise spectrum use, regulatory frameworks need to incorporate dynamic spectrum allocation and sharing procedures.

b. Privacy and data protection: Massive volumes of personal data are gathered by ubiquitous computing through linked devices. To secure consumers' sensitive information, it is imperative that there exist strong data protection laws and privacy rules. Regulators must address important issues including clear permission methods, data anonymisation, and user control over data sharing.

c. Interoperability and standards: Diverse technologies and apparatus are included in heterogeneous networks. In order to minimise vendor lock-in and ensure fair competition, regulatory organisations must promote the implementation of interoperable standards that enable smooth communication across many platforms.

d. Network neutrality: Network neutrality is upheld by regulatory bodies, who make sure that all data traffic is handled similarly. Regulations governing net neutrality prohibit discriminatory actions that could be detrimental to some services or consumers.

e. Sustainability and environmental impact: Regulations may promote the use of energy-efficient technology in network equipment and infrastructure. The negative environmental effects of next-generation networks and ubiquitous computing can be lessened with the aid of green legislation.

f. Emergency services and public safety: Regulatory agencies must make sure that applications for public safety and dependable emergency services are supported by next-generation networks. It is crucial to have standards for network resilience, location precision for emergency calls, and communication during emergencies.

g. Digital divide mitigation: To guarantee equal access to the advantages of ubiquitous computing and next-generation networks, regulators should address the digital divide. Urban and underserved regions can be more effectively connected through policies supporting cheap connection, rural broadband development, and digital literacy efforts.

Collaboration between governments, industry stakeholders, and advocacy organisations is crucial in negotiating these intricate legal and policy problems in order to make sure that new technologies are implemented in a way that benefits society while protecting fundamental rights and values.

12.5.8 Emerging technologies: Quantum computing, terahertz communication

Emerging technologies such as terahertz transmission and quantum computing have the potential to transform heterogeneous next-generation networks and ubiquitous computing, offering up new possibilities for productivity, security, and data processing.

Quantum computing: Quantum computing has great potential for ubiquitous computing and network optimisation due to its ability to do complicated calculations at unmatched speeds. It can tackle issues with encryption, optimisation, and machine learning that are essentially insurmountable for conventional computers. Quantum computing can improve encryption techniques in next-generation networks, providing more reliable data security. Additionally, it can improve data compression, network routing, and resource allocation, resulting in more effective and responsive networks. The ability of quantum computing to quickly solve complex problems might result in advances in materials research, drug development, and artificial intelligence (AI), changing the whole nature of ubiquitous computing.

Terahertz computing: Data transmission rates and bandwidth are unmatched in terahertz communication, which uses a frequency spectrum between microwaves and infrared. Terahertz communication can provide smooth data flow between devices in ubiquitous computing, encouraging real-time interactions and the delivery of high-quality information. Due to its potential for ultra-high-speed wireless connections, it may change how devices interact with one another and enable cutting-edge uses like immediate data exchange and immersive augmented reality experiences. Terahertz communication must find creative methods to overcome obstacles like signal propagation restrictions and atmospheric gas absorption to enable dependable and extensive deployment.

Regulation issues and ethical debates are crucial as these technologies advance. The possible effects of quantum computing on cryptography need revisions to existing encryption standards. Regulation of the spectrum and mitigation of potential privacy and health issues are necessary for terahertz communication. Moreover, it is anticipated that the incorporation of terahertz communication and quantum computing into ubiquitous computing and heterogeneous next-generation networks would accelerate data processing, enable real-time interactions, improve security, and redefine the potential of digital connectedness.

12.5.9 Socio-economic impact and regulatory considerations

Wide-ranging socioeconomic effects result from the combination of ubiquitous computing and heterogeneous next-generation networks, as well as significant regulatory issues.

Socio-economic impact is shown in Figure 12.2, and regulatory considerations are shown in Figure 12.3.

While diverse next-generation networks and ubiquitous computing hold enormous potential for advancement, regulatory frameworks must be

Figure 12.2   Socio-economic impact.

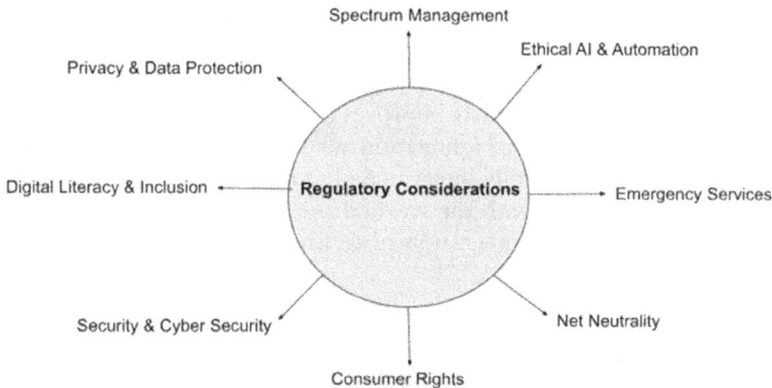

Figure 12.3   Regulatory considerations in socio-economic impact.

forward-looking and address socioeconomic consequences while protecting privacy, security, and ethical concerns. To fully use the potential of new technologies and protect the welfare and values of society, cooperation between governments, industry players, and civil society is essential.

12.6 VISION FOR A FULLY INTEGRATED UBIQUITOUS ECOSYSTEM

A future where technology is smoothly incorporated into every aspect of our life, boosting ease, effectiveness, and wellbeing, is what a fully integrated ubiquitous ecosystem envisions. Devices, apps, and services work together in this vision to build a setting that anticipates and caters to our needs.

This ecosystem's fundamental strength is interconnectivity. Through high-speed and low-latency networks, devices of all kinds – from wearables and smartphones to household appliances and transportation systems – can effortlessly connect with one another. Real-time interactions, personalised experiences, and predictive insights are made possible by the seamless flow of data.

The distinction between the physical and digital worlds is muddled in this ecology. Smart cities use pervasive computing to improve traffic flow, save energy, and provide quick access to public services. Healthcare systems and health monitoring are integrated smoothly, allowing for ongoing health monitoring and quick response. Smart houses change lighting, temperature, and entertainment selections in advance of resident preferences. This ecosystem is driven by artificial intelligence and machine learning, which allow gadgets to adapt to changing environments and learn from human behaviour. Predictive analytics powered by these technologies improve decision-making and enable proactive solutions to new situations.

This perspective does, however, also recognise the crucial significance of ethical issues. Strong protections make sure that user control over personal information is maintained while data privacy and security are of the utmost importance. Individuals are given the tools they need to manage this highly linked world thanks to advances in digital literacy and education. A future where technology harmoniously improves human experiences, enabling efficiency, personalisation, and innovation while upholding basic values and rights is envisioned by a fully integrated ubiquitous ecosystem. The way we live, work, and interact with the world around us is being transformed as a result of this seamless integration of technology, society, and individual wellbeing.

12.7 CONCLUSION

In conclusion, ubiquitous computing has a significant and disruptive impact on heterogeneous next-generation networks (HetNets). A highly linked ecosystem is created by the seamless integration of smart devices and sensors into commonplace items and places, which calls for increased network adaptability and efficiency. HetNets are essential for managing the dynamic connectivity requirements brought on by ubiquitous computing, making use of edge computing and artificial intelligence to maximise network resources and provide customised services. Innovative apps and services appear as the lines between the physical and digital worlds become more hazy, changing user experiences and human-computer interaction. However, the quick spread of linked gadgets also prompts significant worries about data security, privacy, and moral issues. In order to safely utilise the promise of the confluence of ubiquitous computing and HetNets, careful planning, strong security measures, and thorough policy frameworks are required. The combination of ubiquitous computing and HetNets holds the promise of a more connected, intelligent, and inclusive digital future, transforming the way we interact with technology and reshaping our society in unanticipated ways as researchers, policymakers, and industry stakeholders collaborate to address these challenges.

REFERENCES

1. Kameas, Achilles D. "Deploying Ubiquitous Computing Applications on Heterogeneous Next Generation Networks." *IGI Global*, 1 Jan. 1970. www.igi-global.com/chapter/deploying-ubiquitous-computing-applications-heterogeneous/37795
2. "CSDL: IEEE Computer Society." *CSDL | IEEE Computer Society*. www.computer.org/csdl/journal/bd/2022/03/08543852/1DjDlVoyECs   Accessed 13 Aug. 2023
3. Author links open overlay panelMichael Friedewald a, et al. "Ubiquitous Computing: An Overview of Technology Impacts." *Telematics and Informatics*, 22 September 2010. www.sciencedirect.com/science/article/abs/pii/S0736585310000547#:~:text=In%20the%20long%20term%2C%20ubiquitous,and%20in%20the%20medical%20field%2C
4. Ramchand, Anand et al. "The Impact of Ubiquitous Computing Technologies on Business Process Change and Management: The Case of Singapore's National Library Board." *SpringerLink*, 1 Jan. 1970. link.springer.com/chapter/10.1007/0-387-28918-6_12
5. Archetekt. "Ubiquitous Computing: Its Impact on the Workplace and How We Work." *Medium*, 18 October 2018. medium.com/@Archetekt/ubiquitous-computing-its-impact-on-the-workplace-and-how-we-work-5a41944928de
6. Friedewald, Michael. "Ubiquitous Computing: An Overview of Technology Impacts." *Telematics and Informatics*, 15 February. 2015. www.academia.edu/10813024/Ubiquitous_computing_An_overview_of_technology_impacts
7. *Ubiquitous Computing Research: Topics by Science.Gov.* www.science.gov/topicpages/u/ubiquitous+computing+research Accessed 13 Aug. 2023

8. Roe, Sarah. *The Research Center on Values in Emerging Science and Technology*, 17 June 2023. rcvest.southernct.edu/

9. "Heterogeneous Network (Communication System)." *Heterogeneous Network (Communication System)–an Overview | ScienceDirect Topics*. www.sciencedirect.com/topics/engineering/heterogeneous-network-commun ication-system. Accessed 13 August 2023

10. *Network Management in Heterogeneous IOT Networks | IEEE Conference ...*, ieeexplore.ieee.org/document/9498801 Accessed 12 August 2023

11. "What Are Heterogeneous Networks (Hetnets)?" *What Are Heterogeneous Networks (HetNets)?*. blog.3g4g.co.uk/2010/12/what-are-heterogeneous-networks-hetnets.html Accessed 13 August 2023

12. Y. Salsabeel, Shapsough, & Imran A. Zualkernan, Member, IEEE, "A Generic IoT Architecture for Ubiquitous Context-Aware Learning", *IEEE TRANS. On Learning Methodologie*s, Vol. 13, No. 3, pp. 449–464, July–September 2020. doi:10.1109/TLT.2020.3007708

13. Xiuquan Qiao, Schahram Dustdar, Yakun Huang, Junliang Chen, "6G Vision: An AI-Driven Decentralized Network and Service Architecture", *Department: Internet of Things, People, and Processes, IEEE Computer Society*, pp. 33–40, September. 2020. doi:10.1109/MIC.2020.2 987738

14. Mohammed H. Alsharif, Anabi Hilary Kelechi, Mahmoud A. Albreem, "Sixth Generation (6G) Wireless Networks: Vision, Research Activities, Challenges and Potential Solutions," *Symmetry*, Vol. 12, pp. 676, 2020. doi:10.3 390/sym12040676

15. Fengxiao Tang, Yuichi Kawamoto, Nei Kato, and Jiajia Liu, "Future Intelligent and Secure Vehicular Network Toward 6G: MachineLearning Approaches", *Proceedings of the IEEE*, Vol. 108, No. 2, pp. 292–307, February 2020. doi: 10.1109/JPROC.2019.2954595

16. B. Raghothaman, "Architecture and Protocols for LTE Based Device to Device Communications", *Inc. Proc. ICNC*, pp. 895–899, 2013.

For Product Safety Concerns and Information please contact our EU
representative GPSR@taylorandfrancis.com
Taylor & Francis Verlag GmbH, Kaufingerstraße 24, 80331 München, Germany